International Conference on Design & Construction of Super & Mega Yachts 2017

Genoa, Italy
10-11 May 2017

ISBN: 978-1-5108-8301-7

Printed from e-media with permission by:

Curran Associates, Inc.
57 Morehouse Lane
Red Hook, NY 12571

Some format issues inherent in the e-media version may also appear in this print version.

Printed with permission by Curran Associates, Inc. (2020)

For permission requests, please contact The Royal Institution of Naval Architects
at the address below.

The Royal Institution of Naval Architects
8-9 Northumberland Street
London, WC2N 5DA
United Kingdom

Phone: 020 7235 4622
Fax: 020 7259 5912

publications@rina.org.uk

Additional copies of this publication are available from:

Curran Associates, Inc.
57 Morehouse Lane
Red Hook, NY 12571 USA
Phone: 845-758-0400
Fax: 845-758-2633
Email: curran@proceedings.com
Web: www.proceedings.com

CONTENTS

The Royal Institution of Naval Architects

INTERNATIONAL CONFERENCE

Design & Construction of Super & Mega Yachts

10-11 May 2017, Genoa, Italy

INFLUENCE OF DESIGN CHOICES ON SEAKEEPING OF MOTOR YACHTS

F Miguel Montero and **E Della Valentina,** MAritime Research Institute Netherlands (MARIN), The Netherlands

SUMMARY

The objective of this work is to highlight the most important geometrical parameters on heave and pitch motions on displacement mono-hulls. To highlight the influence of the different parameters, frequency domain calculations were performed for a number of existing ships at equal Froude number. The results show a strong dependence of the motions on the ship slenderness and the beam to draft ratio: longer and shallower ships tend to generate more pitch damping, which in turn reduces the motions. Then, systematic variations of a typical hull shape were studied to assess the effect of more specific hull features, such as the waterline entry angle and the submerged stern width. The effects of these changes were always minor compared to the variations observed on the first part of the project, save for extreme variations. The results of this study highlight some of the most important parameters to consider when designing a comfortable displacement motor yacht.

NOMENCLATURE

∇	Displacement (m^3)
Θ	pitch motion amplitude (rad)
ω	Frequency (rad/s)
ω^*	Reduced frequency (-)
A_w	Wave amplitude (m)
B	Beam (m)
CWP	Water plane coefficient (-)
CWPa	Water plane coefficient, aft half of the ship (-)
CWPf	Water plane coefficient, forward half of the ship (-)
Fn	Froude number (-)
g	Gravity acceleration ($m\ s^{-2}$)
Lpp	Length between perpendiculars (m)
LCB	Longitudinal centre of buoyancy, measured from APP (m)
LCF	Longitudinal centre of flotation, measured from APP (m)
Sl	Slenderness (-)
T	Draft (m)
V	Ship speed ($m\ s^{-1}$)
z	Heave motion amplitude (m)

1. INTRODUCTION

A high level of on-board comfort is one of the main objectives when designing a luxury motor yacht. Comfort is the result of a combination of several elements, such as furniture, noise and vibration levels, smell and ship motions. The later are the object of study in the present work. Excessive ship motions can result in sea sickness and loss of postural stability, both of which have a negative impact on the on-board experience.

Decisions made from the very early design stage will have an impact on the seakeeping performance of the vessel. Understanding the consequences of these decisions is therefore of great importance for the designer to design more comfortable yachts.

The present work focuses on the motions in the vertical plane: heave and pitch. Unlike for the roll motion, there are few systems on the market to reduce these motions, and none of them is widespread. Furthermore, heave and pitch motions govern the vertical accelerations, which in turn are the largest contributor to sea sickness. More detailed information on how the vertical accelerations are related to comfort on board can be found in references [1] and [2].

2. OBJECTIVES

The objective of this work is to provide an overview of the main contributing parameters that affect the ship motions and accelerations in the vertical plane and to quantify their influence. This information is meant to be used as guidelines by designers to include seakeeping considerations in the early design stage.

The main parameters considered in the study are the following:

- Main dimensions, in terms of slenderness (defined as $Sl = L\,\nabla^{-\frac{1}{3}}$) and beam-to-draft ratio.
- Bow shape, in terms of waterline entry angle and presence of bulbous bow.
- Aft shape
- Longitudinal inertia

3. METHODOLOGY

Several hull forms were evaluated in head seas and for a range of speeds. The results from these evaluations were compared to illustrate and quantify the differences in pitch and heave motions. A large sample of representative yachts was used to determine the global trends first. On a second stage, specific hull form variations on an average hull form were compared to determine the influence of specific design features.

3.1 CALCULATION METHOD

The motion response of the different vessels were evaluated using a 3D potential flow panel method, FATIMA. FATIMA is a Rankine source based method, where the free surface is modelled explicitly, and allows for the correct modelling of the diffracted wave with forward speed.

Furthermore, FATIMA linearises the problem of evaluating the oscillating wave components with respect to an estimate of the steady flow (including sinkage and trim) and related free surface elevation. This estimate is obtained from the potential flow code RAPID. Apart from these starting points, FATIMA follows traditional linear seakeeping theory by resolving at first the motion induced radiated waves (separately for each motion component, in calm water), the wave induced diffraction pattern and expressing these results in motion induced reaction forces (added mass, damping) and wave induced excitation.

Because the restoring terms are obtained from the steady pressure gradients in RAPID, these deviate from the hydrostatic values used in conventional methods as they account for the actual hull immersion, including sinkage and trim. Combining the foregoing with the location of the centre of gravity, the mass and the inertias of the ship yields the motion response.

A detailed description of the method used in FATIMA can be found in [3].

3.2 IDENTIFICATION OF GLOBAL TRENDS

On the first part of the investigation, a selection of existing hull forms was studied to identify the general trends in pitch and heave motions.

As slenderness is a key parameter in MARIN experience, ships were selected to cover a wide range of slenderness values. A small fishing vessel was included as a very low slenderness reference, and the MARIN Fast Displacement Ship series was included as a high slenderness reference. The characteristics of the selected vessels are summarized in Table 1.

Calculations were run in head seas and for Froude numbers ($Fn = V/\sqrt{gL}$) in the range of 0.15 to 0.25. The longitudinal inertia was set to 25% Lpp in all cases, to eliminate the influence of different weight distributions.

To account for the differences in size, results were compared in terms of non-dimensional RAOs (Response Amplitude Operators). The ship length was used as the reference value for the non-dimensionalisation, as it is the parameter most widely used in the industry to describe yacht size (and, to some extent, cost).

While the RAOs do give a very complete picture of the response characteristics of a single vessel, the trends become more clear when looking at the response in irregular waves. To compare the responses of the different vessels, all the ships were scaled to a length of 80 m (using Froude scaling), and then the RMS (Root Mean Square) values of the heave and pitch motion were calculated for irregular waves of 1 meter significant wave height and with a peak period critical for pitch motion. This period was estimated based on the obtained RAOs, where the peak of pitch was always at a reduced frequency of $\omega^* = \omega\sqrt{L/g} = 0.25$. This is equivalent to a period of about 9 seconds for an 80 m vessel.

3.3 INFLUENCE OF HULL MODIFICATIONS

Geometry variations were performed on a given hull form on a second stage to illustrate the effects of specific hull form features. The hull selected was a typical 80 m motor yacht with bulbous bow shape. The following variations were made on the hull:

- The bulb was removed to assess its influence
- The waterline entry angle and bow flare were made narrowed and wider.
- The aft shape was made narrower and shallower.

In addition to this geometry variations, a case was run with increased and decreased longitudinal inertia. The responses of the different variations were compared in head seas and at a Froude number of 0.2. The motions in head seas were computed using FATIMA for all the

Descriptions	Lpp[m]	B[m]	T [m]	Displ. [ton]	Xcog [m]	Slenderness	B/T	L/B
Heck trawler	30.24	7.20	3.00	332	15.95	4.40	2.40	4.20
39 m tawler yacht	35.90	8.90	2.60	416	16.66	4.85	3.42	4.03
38m expedition yacht	38.08	8.59	2.50	425	18.65	5.11	3.44	4.43
50m expedition yacht	44.67	9.50	3.00	682	19.46	5.12	3.17	4.70
75m motor yacht	66.47	13.62	4.11	2024	29.55	5.30	3.32	4.88
Typical 80m Yacht	72.00	12.40	3.31	1730	34.20	6.05	3.74	5.81
63m motor yacht	62.28	10.50	2.83	1050	27.61	6.18	3.71	5.93
Fast motor yacht	51.90	10.29	2.25	501	24.49	6.59	4.57	5.05
85m slender yacht	83.50	11.82	3.87	1692	37.89	7.07	3.05	7.07
100m CB=0.4 Fast Displacement Ship	100.00	12.50	3.13	1561	44.77	8.69	4.00	8.00

Table 1: Selected vessels for trend identification

above variations and compared to assess the sensitivity of the motions and accelerations.

4. RESULTS AND DISCUSSION

4.1 INFLUENCE OF GLOBAL PARAMETERS

4.1 (a) Heave and pitch motions

The heave and pitch motions show a clear dependence on the slenderness of the vessel and the beam-to-draft ratio. This can be seem in Figures 1 and 2, where the Response Amplitude Operators (RAOs) calculated for the different vessels are presented, coloured (in grey scale) by slenderness and beam-to-draft ratio (only for one Froude number for the sake of space and clarity). To compare vessels of different sizes in a fair manner, the RAOs are presented in non dimensional form (heave over wave height and pitch times ship length over wave height), so that the results are comparable between vessels of different sizes. A clear dependence on both parameters can be seen for both motions, although the effect of the beam-to-draft ratio on the heave motion is relatively limited. In general, it can be seen that shallower and more slender vessels present smaller motions.

The response is a result of the combination of excitation forces, restoring and damping terms. When increasing the slenderness, the excitation forces are increased, but so are the restoring term and the damping. The combination of the three terms results in a decrease of the response. The effect of a shallower hull in terms of excitation and restoring terms is very small, but it does increase the damping. Again, this results in a reduced response. An additional term in the pitch and heave response is the excitation related to the coupling between these two motions. This term depends on the distance between the centre of flotation and the centre of gravity. The effect of this later component will be discussed later with the hull form variations.

In addition to the RAOs, the RMS response in irregular waves was calculated, as this represents a measure of how large will the motions be for a given sea state in a single value that can offer further insight on the trends. Figure 4 shows the RMS values of heave and pitch at Froude number from 0.15 to 0.25, in irregular waves critical for pitch (Peak period chosen such that $w^{*}\sim0.25$) with a significant wave height of 1 m. It can be seen that as the speed increases, the influence of slenderness on the pitch motion is reduced. In any case, once the slenderness is above 6, further increases of the slenderness have limited impact on the response. The heave response always shows a clear trend with the slenderness and it can also be seen that increases with the Froude number, particularly for the low slenderness vessels.

4.1 (b) Vertical accelerations

For a comfort analysis the motions should be interpreted in terms of vertical accelerations, as it is the accelerations that are perceived by the guests and crew on board, rather than the motions. The location on board becomes then a major parameter on the vertical accelerations, which are a function of the longitudinal position on board, the heave and pitch motion and the phasing between them.

Figure 3 (left) shows the distribution of vertical acceleration RMS values along the length of the ship, and highlights the importance of on-board location for a given amenity. The values presented are, once again, scaled to an equivalent length of 80 m using Froude scaling. It can be seen that the accelerations increase rapidly from the point of minimum accelerations with factors up to about 6 at the forward perpendicular. This is of particular interest for the smaller yachts, where the space limitations often result in the need of placing important guests areas on the forward half of the hull.

The previous analysis was done either in a non-dimensional manner or scaling the ships to the same length. To provide an overview of the effects of ship size, the Froude scaling approach was also used to highlight the variations on the perceived accelerations at a fixed on board position with ship length. On the right hand part of Figure 3, the vertical accelerations are calculated at 75% Lpp for all vessels and applying Froude scaling to a range of ship lengths. Because the response in irregular waves is strongly dependent on the

	Displacement [m³]	LCB [m]	LCF [m]	(LCB-LCF)/Lpp [%]	CWP	CWPa	CWPf	Slenderness
Original hull	1685	34.16	29.10	7.03	0.812	0.971	0.652	6.05
No bulb	1649	***33.37***	***29.11***	***5.92***	0.812	0.971	0.652	6.09
Wider bow	1692	***34.29***	***29.40***	***6.79***	***0.819***	0.971	***0.667***	6.04
Narrower bow	1675	***34.02***	***28.70***	***7.39***	***0.802***	0.971	***0.631***	6.06
Narrower stern 1	1673	***34.38***	***29.45***	***6.85***	***0.803***	***0.952***	0.652	6.07
Narrower stern 2	1652	***34.77***	***30.21***	***6.33***	***0.781***	***0.911***	0.652	6.09
Narrower stern 3	1604	***35.59***	***32.29***	***4.58***	***0.750***	***0.859***	0.652	6.15

Table 2: Hull form variations on the typical 80 m motor yacht

peak period, the period was also scaled so that it was always critical for pitch (keeping $\omega^* = 0.25$). In this case, the accelerations are significantly reduced for the longer ships, as they are inversely proportional to the length.

4.2 INFLUENCE OF SPECIFIC DESIGN FEATURES

Systematic variations were performed on the "Typical 80 m motor yacht", as presented on Table 2, to assess the influence of certain design modifications on the heave and pitch response. The following subsections describe and discuss the different variations. Except for when explicitly mentioned, all the calculations were done assuming a radius of longitudinal inertia equal to 25% of the ship length.

4.2 (a) Influence of the bulb

Figure 5 shows the heave and pitch RAOs for the hull with bulb (original shape) and without bulb. The influence in the pitch motion is negligible, but the variation with no bulb showed slightly less heave. A likely explanation for this is that the calculations were always performed in even keel condition, and thus the centre of gravity (COG) was moved to match the longitudinal centre of buoyancy (LCB), that is shifted because of the lack of bulb. The change in distance from the centre of the flotation (LCB) to the centre of gravity results in a reduced heave-pitch coupling term. This effect is noticeable on the heave and not on the pitch because, for the frequencies near the maximum pitch, the wave excitation forces for pitch are much larger than for heave.

In terms of vertical accelerations along the hull length it can be seen that the minimum value is lower than for the reference vessel, and then the slope is roughly parallel, which is a result of the pitch motion being practically the same. This can be seen in Figure 9. A result of this is that the difference of accelerations is likely to only be noticeable in the amidships region, but the impact there will be limited as this is a region where the accelerations are low anyway.

4.2 (b) Influence of the fore body shape

On the basis of previous MARIN experience, it was expected that the waterline entry angle and bow flare had an effect in pitch damping. This was checked by making a hull form with a narrower and wider water line entry angle and bow flare. The results for this study can be seen in Figure 6. The results show that a wider bow slightly reduces pitch, but the main difference comes in the heave motion. This is probably the result of the larger pitch in combination with a larger coupling term.

An important remark here is that the effects listed on this section are only caused by the wider waterline. The more or less pronounced bow flare above the waterline is not considered in the linear calculations other than the small effect it has on the steady state wave train.

The effects of these variations in terms of vertical accelerations can be noticed forward of station 7, as can be seen in Figure 9. The vertical acceleration RMS at station 10, for instance, shows an increase of about 5% for the narrower waterline entry version, and a decrease of 3% for the wider waterline. This represents a limited, but not negligible impact on the on board comfort.

4.2 (c) Influence of the aft body shape

In MARIN's experience, the cruiser stern shape results in larger motions on the vertical plane. To assess the extent of this effect, the stern of the reference vessel was made narrower and rounder in 3 steps. The calculated RAOs can be seen in Figure 7. In general, the impact of these modifications in heave and pitch is relatively limited until the last step, where at the waterline the stern is more rounded and the waterline length is reduced.

The heave and pitch are increased when the stern is made narrower. This is a result of the lower damping generated by the stern. The coupling between heave and pitch is also reduced, but the increase in pitch still results in an increase in heave. A consequence of the reduced LCB to LCF distance is that the point of minimum accelerations is shifted forward.

In Figure 9 it can be seen that the impact of the stern modifications in terms of accelerations is somewhat smaller than that of the fore body modifications for the motions in front; save for the very extreme third step. In the aft, however, the accelerations are noticeably higher, as a result of the higher pitch and centre of motions being moved forward.

4.2 (d) Influence of the longitudinal inertia

The longitudinal inertia of the vessel represents an important term in the equations of motion that may have equal or greater impact on the response than the hull geometry. In order to assess the sensitivity to this parameter, the reference ship calculations were run with different radii of inertia (24% and 26%). The obtained heave and pitch RAOs can be seen in Figure 8. An increase of the motions with the radius of inertia can be appreciated.

While the impact of small changes of the radius of inertia such as those used here is relatively small, it is still on the same order of magnitude as the changes in hull geometry. This can be seen more clearly when comparing the acceleration RMS in Figure 9, where the impact of the inertia changes is similar to the fore body modifications.

5. CONCLUSIONS

The results of the investigation on the motions in the vertical plane of a collection of vessels and of systematic variations of a typical hull form suggest the following conclusions:

The slenderness and beam-to-draft ratio of a given vessel largely condition the response in head seas. In general, a more slender and shallower hull presents smaller motions. However, once these parameters are high enough, the impact of further increasing them is relatively small.

At higher ship speeds, the heave response increases, particularly for ships with low slenderness.

No matter the general characteristics of the vessel, the on-bard experience may change significantly from one location to another. This is a particularly important consideration when designing the general arrangement of a motor yacht.

The impact of the bulb on the vertical motions is very limited, but the different position of the COG required to keep an even keel may affect the heave-pitch coupling. Fuller fore and aft bodies reduce the motions on the vertical plane. The effect of relatively small variations of the longitudinal weight distribution can be compared to those of changing the hull geometry.

The impact of the studied modifications is in any case limited and smaller than the variability seen between different designs. It is therefore of great importance to take the ship motions into consideration from the beginning of the design process, as it is likely to be difficult to have a significant impact on these by minor changes on the project.

6. REFERENCES

1. DALLINGA, R.P., BOS, J.E., 'Cruise ship seakeeping and passenger comfort', *Int.Conf. on Human Performance at Sea*, Glasgow, 2010

2. EU Compass Project G3RD-CT-2002-00809 Final Publishable Report "A rational approach for reduction of motion sickness & improvement of passenger comfort & safety in sea transportation"

3. BUNNIK, T., 'Seakeeping Calculations for Ships, taking into account the non-linear steady Waves', *PhD Thesis, Delft University of Technology*

7. AUTHORS BIOGRAPHY

Francisco Miguel Montero *(f.miguel@marin.nl)* holds the current position of Project Manager on the yachts market team at MARIN. He is responsible for seakeeping studies, by both numerical methods and model tests. He is a naval architect by the Universidad Politecnica de Madrid (Spain) and has a Masters in Naval Architecture by the University of Michigan (USA). Professionally, he has over 6 years of experience with seakeeping projects, both for ships and floating structures.

Enrico Della Valentina *(e.valentina@marin.nl)* leads the yachts market team at the Ships department of MARIN. With a double Master in Naval Architecture, one from the Universita' degli Studi di Trieste (Italy) and one from the Technical University of Delft (The Netherlands), he has been working at MARIN for 13 years managing projects initially in the field of manoeuvring, then in the field of powering and later in seakeeping. Multidisciplinary projects are his specialty.

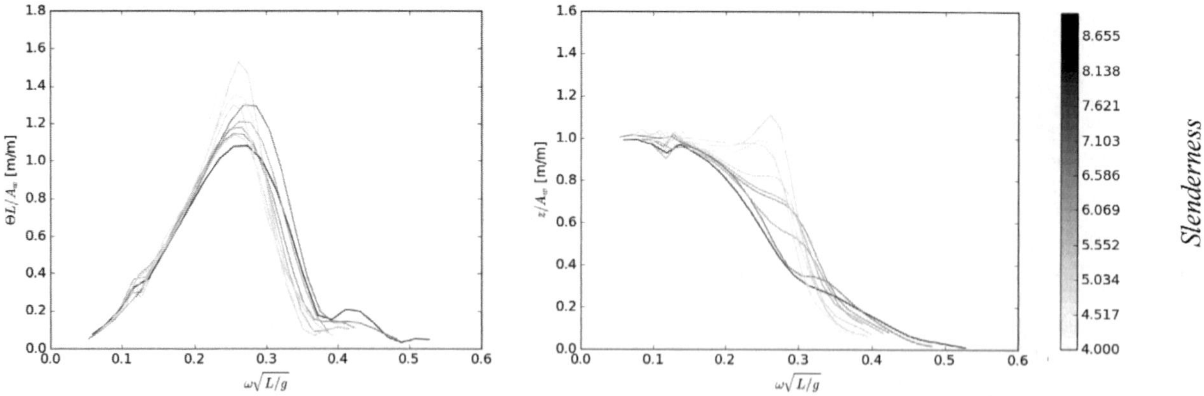

Figure 1 Non-dimensional pitch (left) and heave(right) RAOs at Fn=0.2, colours by ship slenderness

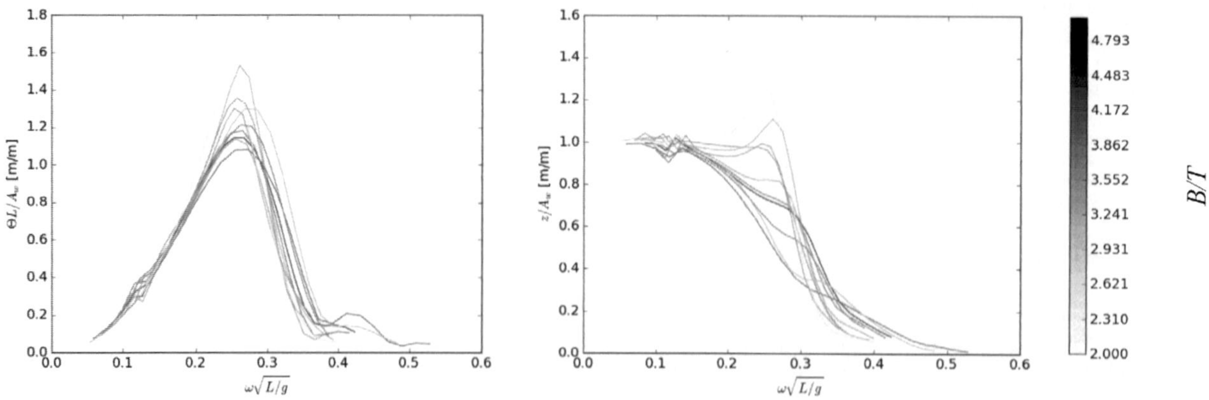

Figure 2 Non-dimensional pitch (left) and heave(right) RAOs at Fn=0.2, colours by ship beam-to-draft ratio

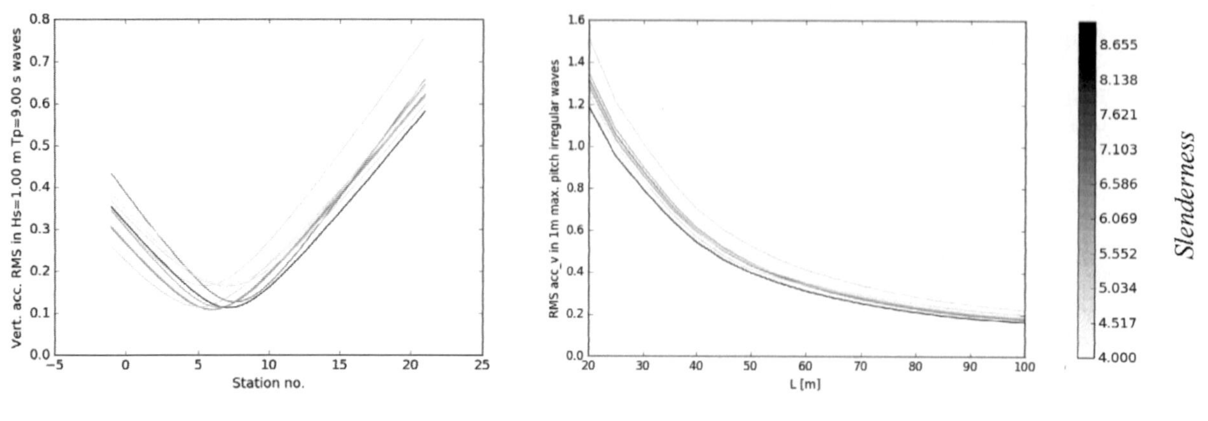

Figure 3 LEFT: Vertical acceleration RMS values (in m/s^2) along the ship length at Fn=0.2, colours by ship slenderness, all ships scaled to 80 m. RIGHT Influence of ship size on vertical acceleration RMS values (in m/s^2) at 75% Lpp

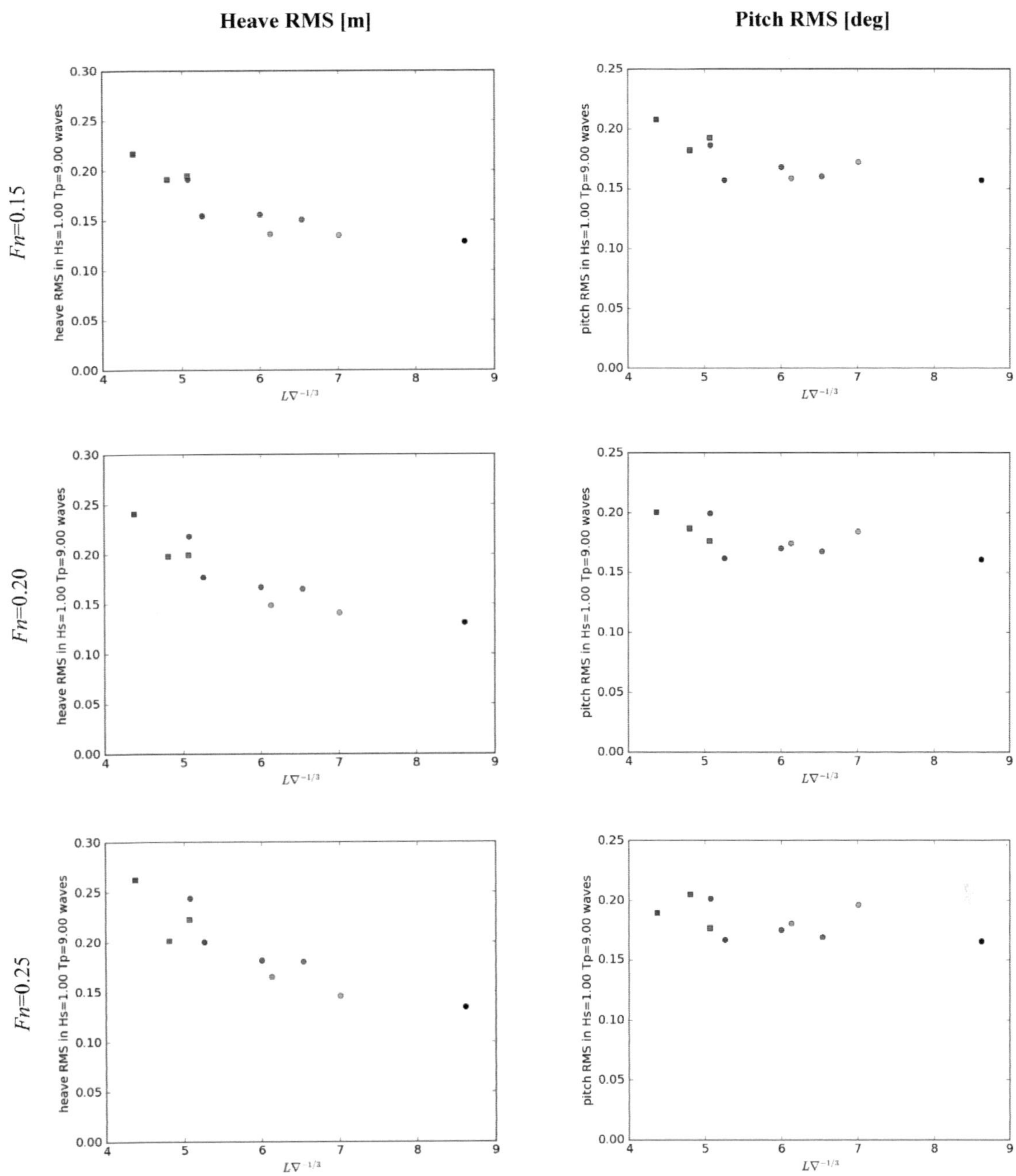

Figure 4: Heave and pitch response for all vessels (Froude scaled to 80 m ship length) as a function of slenderness, for Froude numbers of 0.15, 0.20 and 0.2; in irregular waves critical for pitch motion: Hs=1 m and Tp=9.0 s (JONSWAP spectrum). The different markers represent each of the studied vessels.

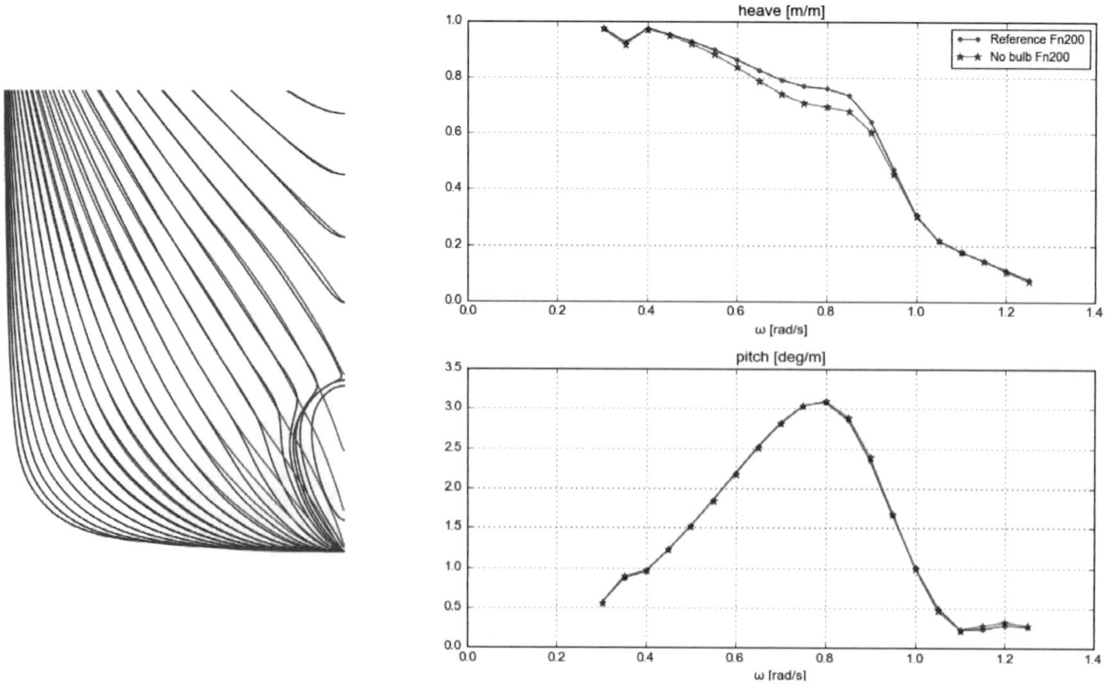

Figure 5: Influence of the bulb on the heave and pitch response

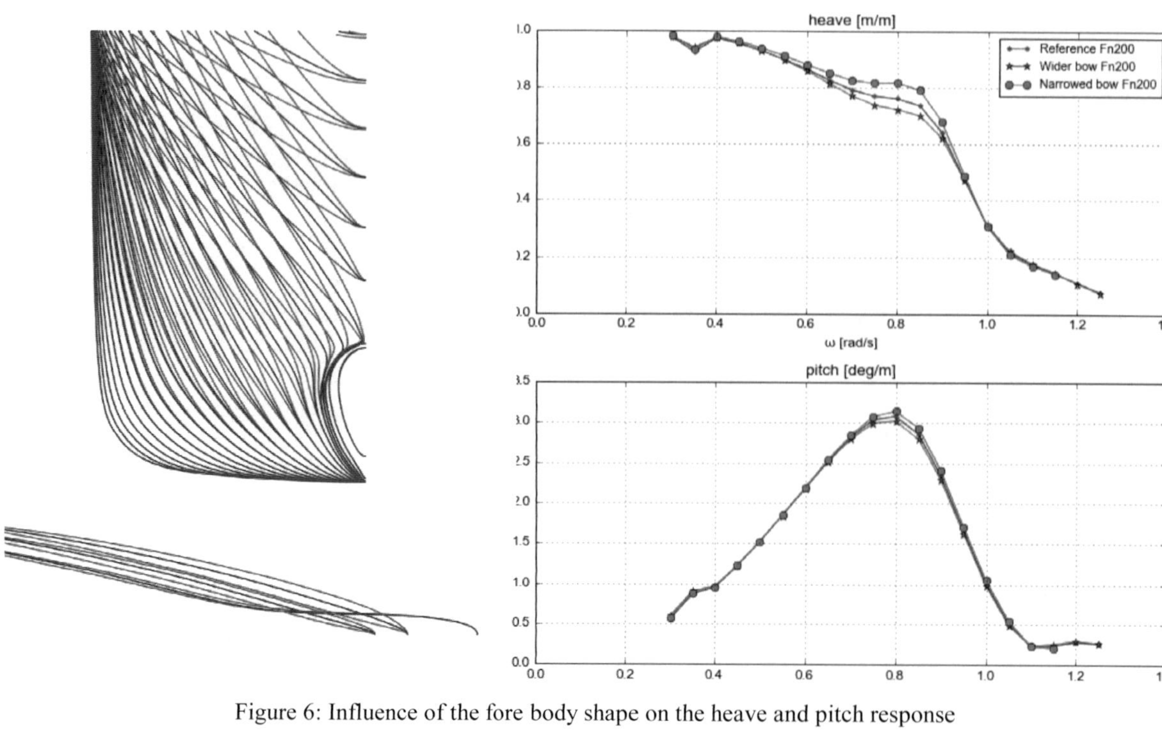

Figure 6: Influence of the fore body shape on the heave and pitch response

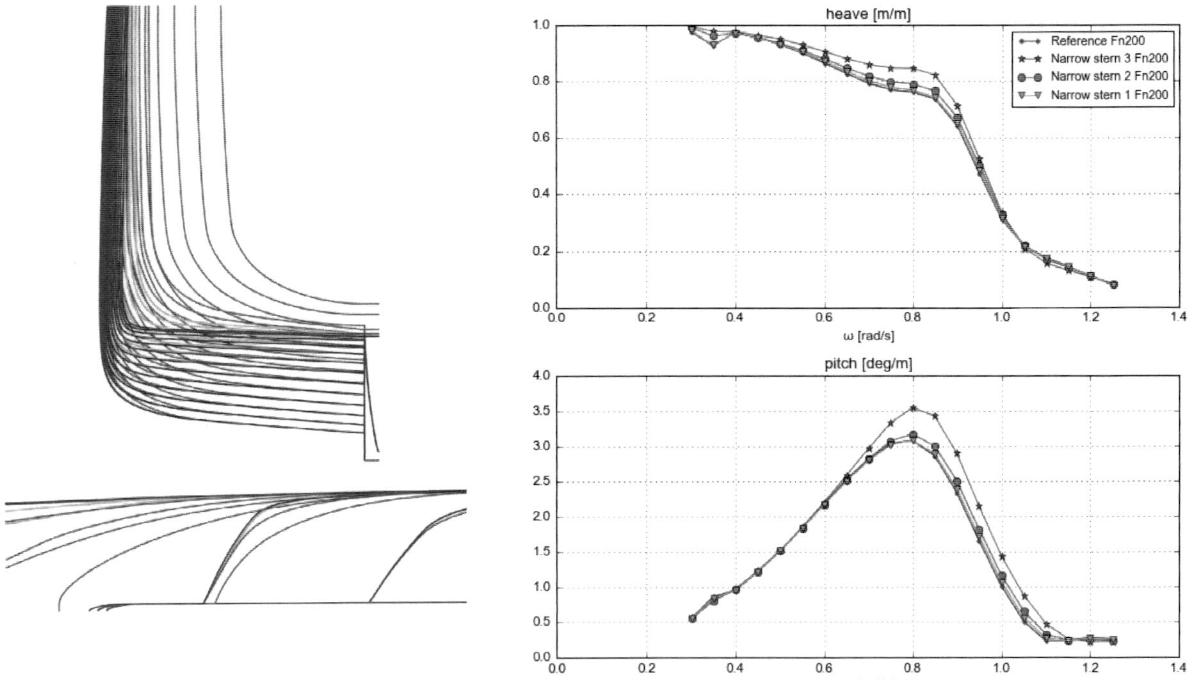

Figure 7: Influence of the aft body shape on the heave and pitch response

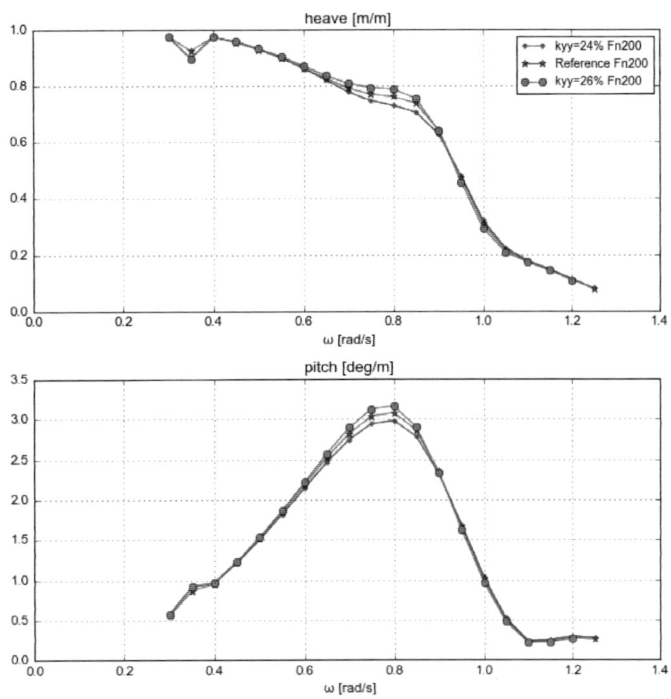

Figure 8: Influence of the longitudinal inertia on the heave and pitch response

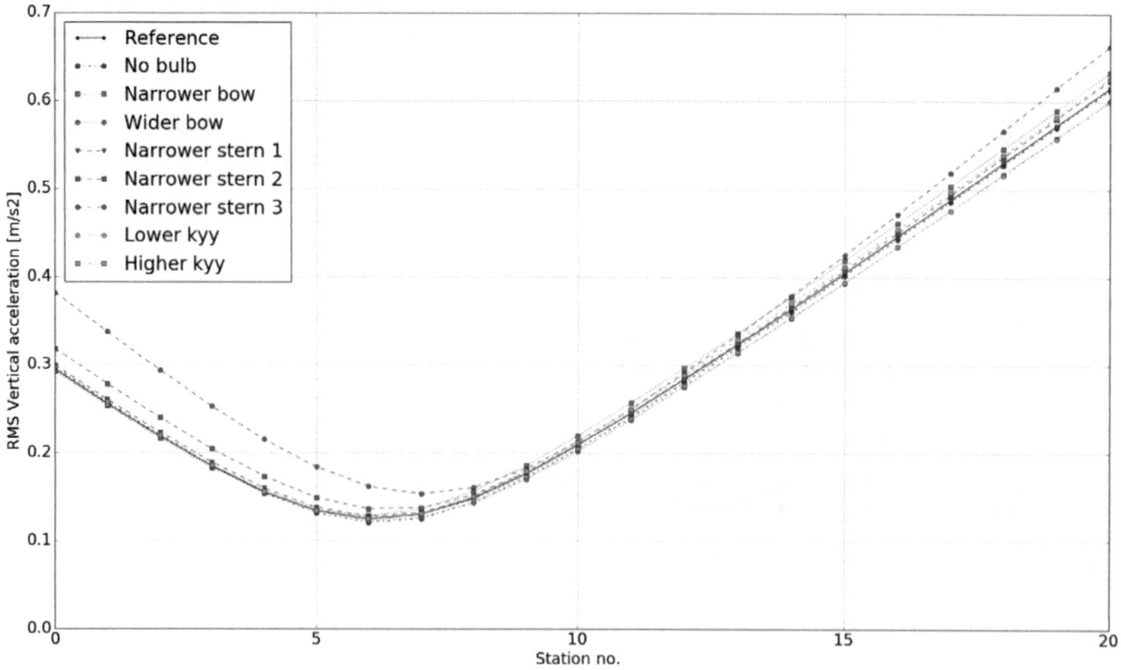

Figure 9: Vertical accelerations RMS is irregular waves of Hs=1 m and Tp=9.0 s, for the different variations of the typical 80 m motor yacht.

MANAGING THE MEGA-YACHT'S WASTEWATER STREAM

M. A. Beavis, ACO Marine Systems, Germany

SUMMARY

IMO Resolution MEPC 227(64) has changed the rules governing the use of on-board wastewater treatment plant and impacts the design and construction of mega-yachts, as the revised regulation makes some existing plant incompatible. This paper will outline the impact of MEPC 227(64) for larger PYC yachts with more than 12 passengers and discusses the limitations of various technologies and the design and technical implications of under-specifying wastewater treatment plant. Referring to case studies involving the installation of wastewater systems aboard a number of high profile mega-yacht projects, the paper will explain how designers can reduce the space required for water treatment plant without impacting treatment capacity and performance.

The paper will go on to explore the technical challenges faced when ACO Marine designed a wastewater treatment system to meet the limited, very narrow space aboard one particular prestigious project.

NOMENCLATURE

AWTP	Advanced Wastewater Treatment Plant
BOD	Biological Oxygen Demand (mg/l)
COD	Chemical Oxygen Demand (mg/l)
FOGs	Fats, Oils and Greases
LY3	Large Yacht Code 3 [up to 12 passengers]
MBR	Membrane Bio-Reactor
PLC	Programmable Logic Controller
PYC	Passenger Yacht Code [>12 passengers]
STP	Sewage Treatment Plant
TSS	Total Suspended Solids
UV	Ultra Violet

1. INTRODUCTION

Although in essence a 'mission-critical' system, in reality wastewater treatment is so often an afterthought for both designers and naval architects involved in the design and construction of arguably the world's most prestigious and innovative vessels. Many systems onboard can technically malfunction without passengers and staff ever noticing ……… but if the wastewater system associated with the bathrooms and heads of the vessel fails, it will be noticed by everyone on the vessel. Similarly, if the treatment process is incorrectly specified a process overload or even breakdown could result in pollution and environmental breaches often attracting hefty fines and, in extreme cases, even criminal proceedings against the crew and owners of the vessel.

But this doomsday scenario can be avoided if consideration is given to the development of the wastewater management philosophy at the earliest possible stage of the ship design. Time and space provisioned to the development of this element of the yacht's operation will yield considerable dividends later during the detailed design and construction phases of the build. Important is not just to look at the treatment aspect of the problem, but to look at the problem from a 'whole ship' perspective. To understand what wastewater is being produced where, in what quantities, at what frequency and how it will be stored and transferred around the vessel prior to, and after, treatment by the AWTP.

2. UNDERSTANDING IMO MEPC 227(64)

On 1st January 2016, Resolution MEPC 227(64) [1], adopted by the Marine Environmental Protection Committee of the International Maritime Organisation in October 2012, came into force. The amendment changed previous rules governing the testing, performance and operation of on-board wastewater treatment plant and will impact heavily on the design and operation of super and mega yachts.

Given the increasing trend for expedition-type yachting in ecologically sensitive areas such as the Baltic, the new regulation will make some existing plant incompatible as MEPC 227(64) [1] changes both discharge requirements and test protocols. It has the specific aim of reducing nitrogen and phosphorous from passenger vessel treated water discharges, preventing the acceleration of nitrification of the seas.

For PYC vessels (with more than 12 passengers) the Section 4.2 of the revision specifically prohibits black (sewage) water discharge into the Baltic Sea special area, with the exception of vessels equipped with a type-approved wastewater treatment system that meets effluent discharge requirements of less that 10mg/l of Nitrogen and less than 1.0mg/l of Phosphorous.

'Special Area' designation and clarification of "Passenger Vessel" definition are determined and detailed within the IMO MEPC 200(62) [2] resolution. Currently the only IMO designated special area is the Baltic Sea, but more areas around the world are in the process of applying for similar recognition.

Furthermore, according to a survey carried out in 2012, one IMO Member State found that a significant number of treatment systems did not meet the existing standards, due to improper use of detergent, a lack of maintenance, or simply because operators failed to follow the plant manufacturer's operating instructions.

But IMO MEPC 227(64) [1] also applies to non-passenger (LY3) yachts globally as this new resolution supersedes IMO MEPC 159(55) [3] and currently there are only a limited number of wastewater treatment plant technologies on the market that meet the new requirements. This is largely due to the fact that systems that have hitherto relied on dilution as a part of the treatment process will have difficulty in meeting the effluent standards and test protocols of the revised Resolution.

The requirements, with the exception of the requirements in section 4.2, will apply to sewage treatment plants installed on or after 1st January 2016 on: ships, other than passenger ships, in all areas; and passenger ships outside MARPOL Annex IV special areas.

The Section 4.2 of these Guidelines for passenger ships will apply to sewage treatment plants installed on:

- New passenger ships when operating in a MARPOL Annex IV special area and intending to discharge treated sewage effluent into the sea on or after 1st January 2019; and [amended to June 2019]

- Existing passenger ships when operating in a MARPOL Annex IV special area and intending to discharge treated sewage effluent into the sea on or after 1st January 2021 [amended to June 2021].

It should also be noted that, when ships are operating approved sewage treatment plants, MARPOL Annex IV also provides that the effluent shall not produce visible floating solids or cause discolouration of the surrounding water."

MEPC 227(64) also states that an approved sewage treatment plant should not rely solely on dilution of wastewater. And where amounts of dilution are deemed essential to a given treatment process/technology, effluent standards have to meet more stringent criteria.

3. TREATMENT TECHNOLOGIES

3.1 IS DILUTION REALLY A SOLUTION?

No.

Many technologies in use on larger yachts have, until now, used technologies that depend on dilution and chemical sterilisation to achieve permitted effluent standards due, mainly, to their perceived smaller space requirement and lower initial purchase cost. But "dilution is not a solution to pollution"; it is not treating the waste, it is merely 'reshaping' it. Treatment technologies have now to be seen to effectively treat and remove the pollutants, not just 'dilute and dump'.

Other systems utilise a combination of filtration and sterilisation by chemical dosing, electro-catalytic action and even in one case the use of Hydrogen Peroxide. These are not 'environmentally friendly' solutions. Chemical treatment relies on the storage and use on board of chemicals with limited shelf life and transport/storage difficulties. Some are also listed as hazardous materials. Electro-Catalytic systems require circulation of salt water to produce chlorine which can be problematic in fresh water locations and dangerous if not maintained and operated correctly. Filtration stages require frequent maintenance and sludge generation rates are very high. Hydrogen Peroxide is highly corrosive and can be particularly dangerous if not handled properly.

For these reasons the predominant technologies installed on larger yachts utilise some form of biological process as this is effectively taking a natural process and, by the use of extended aeration, 'super-charging' the bacteria to do what they do best, namely to eat the waste. The resulting inorganic, or surplus sludge, is made up primarily of the dead bacteria skeletons and other inorganic material which can be safely discharged to shore, or to sea (local rules permitting).

3.2 ADVANCED MEMBRANE BIOREACTOR TREATMENT

These are the systems most commonly found on specialist vessels such as yachts and other passenger ships where a truly flexible and global operating profile is anticipated. Additionally some vessel types, under Classification Society Green Notations, are also using MBR solutions.

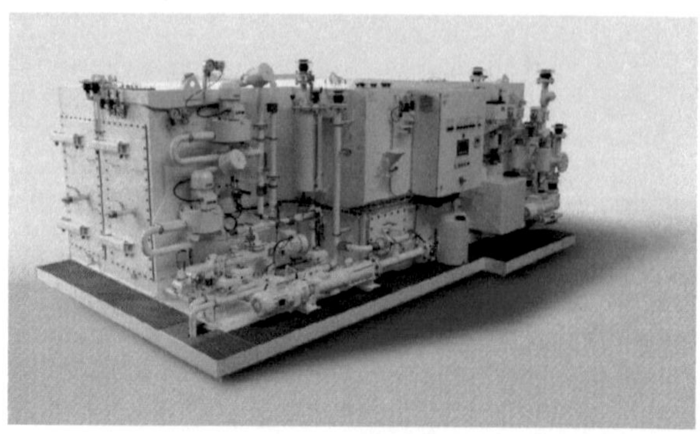

Figure 1: ACO Marine's Maripur NF is the wastewater solution of choice for PYC yacht owners and builders

Whilst often considered to be highly complex and labour intensive to operate, developments in membrane material technology as well as PLC control enhancements now mean that membrane solutions form the back bone of wastewater management solutions on most types of specialist and passenger vessels. Their reliability and treatment performance are proven and regarded by many as the industry standard.

Irrespective of the membrane technology used, submerged flat plate or external tubular, the cross flow membrane principle provides a physical barrier ensuring that only water that meets the IMO effluent standard is able to pass through the membrane and subsequent discharge overboard.

Membrane effectiveness is such that no sterilisation chemicals or UV are required. Whilst the clear effluent is not considered potable, the treated water can be directed for re-use as technical fresh water for toilet flushing, deck washing or laundry applications. Alternatively it can be discharged overboard without reproach by the authorities, even in IMO Special Areas.

3.3 ENHANCED CONVENTIONAL BIOLOGICAL TREATMENT

Designed by ACO Marine and offered as an alternative to AWTP options, these offer a simpler and more cost effective solution for LY3 category yachts (no more than 12 passengers) where the compliance with Section 4.2 Nitrogen and Phosphorous removal is not required. These are now fast becoming the industry standard for the 50 metre plus LY3 yachts and recognised as a viable and practical alternative to physical chemical and dilution technologies. Although larger than their chemical counterparts, they usually do not require use of ship's buffering tanks and so the overall space requirement for the whole system is comparable in overall system size.

Figure 2: ACO Marine's Clarimar MF

Innovative 'bio-sword' filtration technologies eliminate the requirement for the conventional settling stage of the process thereby nullifying any effect of ships motion and at the same time again providing a physical barrier ensuring only IMO compliant discharges pass overboard. The unique properties of the ACO "bio-sword" allow operation with bio-mass concentrations in the activation chamber up to four times higher than those of conventional settling type sewage treatment plants. By operating at much higher concentrations a greatly reduced volume of activation tank is achievable with an overall significant reduction in both the footprint and maintenance envelope requirement. This makes the bio-sword solution one of, if not the, smallest biological treatment units on the market.

No chemicals are required for the process with sterilisation of the effluent water being achieved by the use of in-line UV sterilisation unit(s). The new technology reduces pollutants to well below the mandatory requirement. For instance, the requirements stipulate that the maximum allowable Coliform Bacteria discharge is 100/100ml, but tests have shown that both the ACO Clarimar MF and ACO Maripur NF systems can reduce this to 25/100ml. TSS content in the effluent, mandated not to exceed 35mg/l, can be reduced to <14mg/l; BOD at 25mg/l can be treated to <3mg/l; and COD to <25mg/l, 100mg/l less than the mandatory requirement. Acidity/alkalinity levels are now mandated to be within the range 6-8.5pH. The ACO Marine plants have a Type Approval pH value of 7.44.

Chlorine is neither used nor produced.

3.4 CONVENTIONAL SETTLING TYPE BIOLOGICAL TREATMENT

Whilst an effective and historically well-proven system the technology is dated now and generally considered to be too large for practical use on any class of yacht.

The settling stage is susceptible to ship movement and as the process works on the flow through principle cascading from one stage to the next, any hydraulic overload will simply flush the wastewater directly overboard. This technology requires the use of chlorine disinfection and de-chlorination injection for sterilisation as the treated effluent water is generally too 'coloured' for ultra violet to be effective.

4. CASE STUDIES

4.1 96 METRE SUPERYACHT

Problem No.1 – there is never enough space in a yacht, no matter how big it is.

In 2007 ACO Marine was asked to develop a solution that would provide full 100% treatment redundancy with two units each capable of taking the vessels full load. However the space provisioned did not take into account the redundancy requirement that was added to the specification part way through the project.

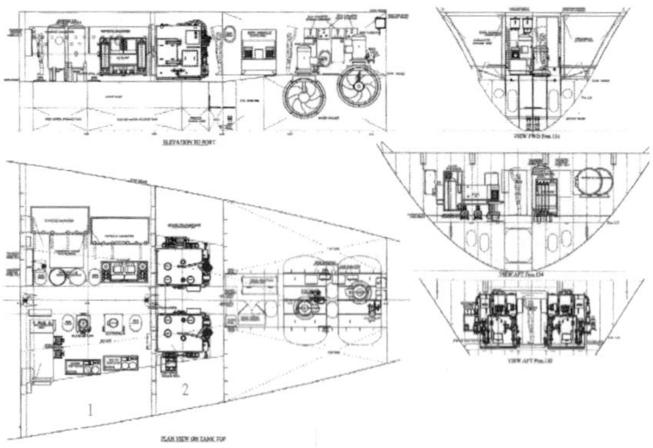

Figure 3: 'Mirrored' units and repositioned access covers allow installation in smaller compartment

The shipyard originally planned to install two MBR units in a forward compartment – labelled No.1 in Figure 3 above. Modelling of the compartments, together with flexible design innovation, allowed two Maripur AM100 units to be placed into the smaller neighbouring compartment – labelled No.2 in Figure 3 above. The latter compartment had initially been crammed full with calorifiers and other equipment which was then able to be relocated into compartment No.1 thus freeing up space and simplifying access to this unrelated equipment.

Treatment units were configured so that all operation maintenance tasks could be accessed along the centre line of the compartment and the end walls of the units kept blank to allow installation tight against the compartment's aft bulkhead. Access covers on the top of the units were repositioned to ensure they fell between the deck-head deep frames to allow maximum use of available height and still allow access for inspection.

This was an example of ACO Marine working together with the yard to solve not only the wastewater treatment issues on board but also to provide a flexible solution that contributed toward solving other space-related issues elsewhere in the vessel.

4.2 146 METRE MEGAYACHT

Problem No.2 – there is still never enough space in a yacht. The forgotten equipment.

In late 2015 into early 2016 ACO Marine supplied and delivered this project which, although a newbuild vessel, presented much like a retrofit problem. The only space left available for the wastewater system, its transfer components, grease separation and black water vacuum collection station was a long thin compartment aft in the vessel to the port side.

One problem with an installation in the aft section of a vessel is the tapering hull form combined with the routing of main engine propeller shafts. The port engine shaft passed along the full length of the inboard side of the wastewater treatment room.

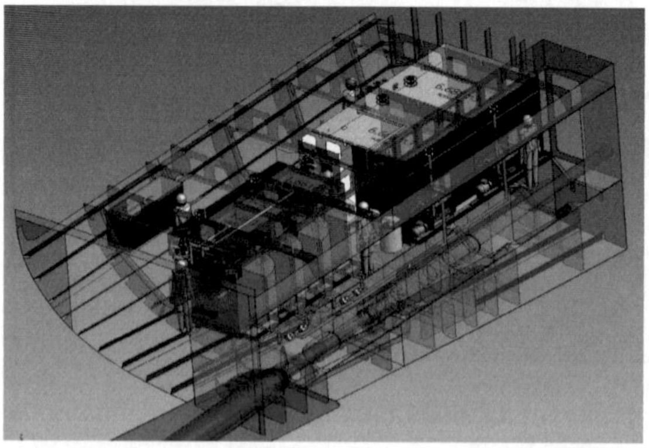

Figure 4: Screenshot of in-line treatment solution using 3D compartment modelling

To meet the challenge the first and second stage water treatment units were designed and configured end-to-end requiring a more complex integration into the space and with the ship systems. Cross connection pipes between stages had to be sloped to provide a fall (sloping aft) for the cross-connection of the stages. All openings and maintenance spaces were arranged facing either fwd/aft or inboard to open over the shaft bed plates. The outboard sides of the treatment units were kept blank to allow installation as close to the ship's side as possible. This meant that the main engine shaft could still be freely accessed to allow shaft and shaft bearing maintenance and/or removal as required.

The treatment plants also had to be internally subdivided to offer full 100% system redundancy.

Another space-saving example is the back-to-back installation of 'mirrored' units to minimise maintenance space requirement. An example of such a solution on a 152 metre megayacht can be seen in Figure 5 below.

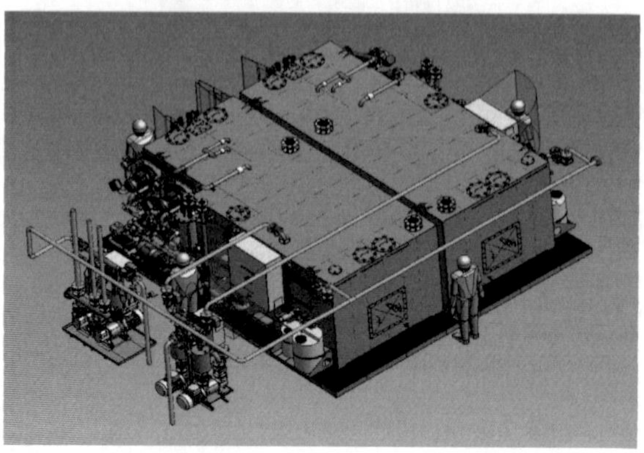

Figure 5: Mirrored back-to-back units reduce the overall maintenance envelope

4.3 SUPER & MEGA YACHTS IN GENERAL

Problem No.3 – (apart from never having enough space). Extended periods of vessel 'down time'.

Biological treatment processes rely upon the bacteria within the wastewater influent to digest the waste material contained therein. However, as well as considering overload scenarios, often overlooked can be the potential for under-loading. This can be equally challenging for wastewater designers and so ACO Marine developed its central wastewater management system, known simply as RM3, which provides centralised monitoring and control of all ACO supplied components within the system and their interaction with ship's tanks and systems to ensure controlled transfer from the various collecting tanks around the vessel to a central mixing tank from which a controlled transfer to the treatment unit is achieved.

Part of the RM3 design includes what is referred to as '2-in-1' functionality. Usually large yachts will have two treatment plants installed to meet the full load demand but with one unit being shut down during extended periods of downtime to maintain reasonable biological concentrations within the process. When passengers come on board the off-duty unit is brought back on line by desludging the running unit into the standby unit. This process is referred to as 'cross-seeding'.

This all works very effectively but does mean that two units have to be installed at a higher cost both financially and in terms of space. It is also an additional procedure required to be carried out by the crew.

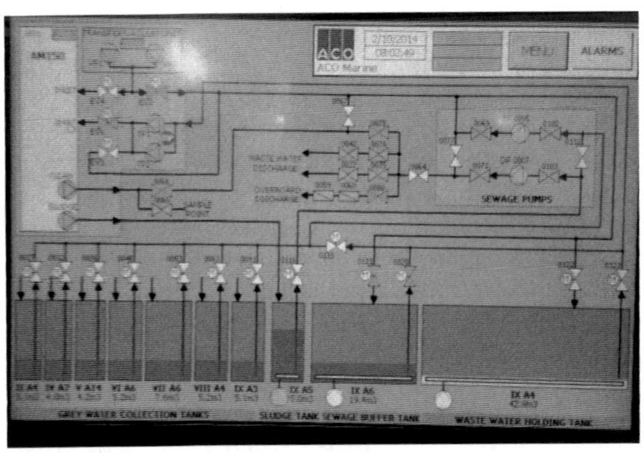

Figure 6: RM3 example touch screen display

To simplify the operational transition from 'crew only' manning to full load manning, the 2-in-1 concept means that, on ships where conditions are suitable, one large MBR unit can be installed and several loading scenarios programmed into the RM3 system's PLC. Typical settings include, but are not limited to, *crew only / crew & guests / 'party'* mode. By simply making the selection on the RM3 panel's touch screen menu the appropriate mode can be implemented. The system then automatically adjusts the operating parameters of the process to facilitate varying load scenarios by reducing operating levels and cycling membrane elements so that only the optimal number of membrane modules are on line at any given time.

This feature minimises system space requirements, provides almost instantaneous transition between differing loading scenarios and can all be implemented from one central point.

5. THE IMPORTANCE OF EFFECTIVE GREASE MANAGEMENT

The effective and reliable treatment of ship's galley water by the removal of waste solids (sludge) along with fats, oils and greases (FOGs) is vital to ensure the effective operation of the downstream wastewater treatment system. Failure to effectively separate and remove the galley FOGs and sludge will cause biological overloading of any treatment system and promote system blockages.

Whilst there are currently no IMO rules regarding the standards of grease separation from galley water, once it enters the wastewater treatment system it must meet the IMO MEPC 227(64) requirement for sewage discharge which does not allow the discharge of any oils. Discussions are on-going in various legislative quarters to bring in requirements for the treatment standards of galley water that is directly discharged overboard, so simply bypassing the treatment system will almost certainly not be a solution in the longer term.

The important word is "separator". Earlier pre-conceived notions that grease "traps" were sufficient have since been proven incorrect. These are traditionally small, in-line, entrapment boxes fitting the galley with nothing more than a weir to hold back floating grease and oils. It takes no account of neutrally buoyant substances or more dense solids and is prone to carry-over of the greases and oils onto the discharge side of the unit. These units require opening, manual emptying and cleaning on a daily basis. As this is a particularly unpleasant task, as well as a potential health hazard, it is often not undertaken frequently enough; or even not carried out at all. Therefore the only viable solution to ensure reliability of the wastewater treatment system as a whole is to fit fully enclosed self-cleaning grease separation technology in a yacht's machinery space where it can be maintained by ship's engineers in the same manner as all other machinery.

5.1 GREASE SEPARATOR DESIGN

To meet the requirements of operation on board ships, separators should be totally enclosed to avoid any possibility of odours into the compartment and fully flooded to remove any free surface effect caused by vessel movement that could lead to re-mixing of the grease/oil into an emulsion which is more difficult to separate.

ACO Marine Lipator and Lipatomat Grease Separators are self-cleaning systems that allow continued on-line availability should be considered to ensure zero down-time of the plant and continuous treatment 24 hours a day. Self-cleaning configurations remove the requirement for regular operator intervention and ensure optimum performance.

The shape of separators should be such to encourage efficient separation whilst at the same time having 'clean' internal surfaces that do not allow build-up of solids behind frames or in corners. The most effective design is a cylindrical concept with cones mounted top and bottom for effective channelling of FOGs and solids to the discharge points.

5.2 GREASE SEPARATOR SIZING

In the interests of reducing costs and minimising space requirement, grease separator sizing is often not carried out correctly. Consideration should not only be given to the number of persons on board, but also to additional, equally important, factors such as sporadic loading, detergent content, water temperature, operational periods of the galley, density of the galley water as well as hydraulic load per meal. It is therefore vitally important that, as with the MBR selection, when developing solutions for galley water treatment the operational profile of the vessel should be fully understood.

Of the entire 'hotel' (toilets, sanitary, laundry and galley) wastewater generated onboard ships, galley water is the most difficult to manage and effectively treat. Yet it is the one that is most often overlooked. Fats, oils and greases are the single biggest contributing factor to wastewater treatment system failure, regardless of the treatment technology used. Early provision in the ship design for the correct grease separation technology will ensure reliable operation, and lowest risk, to the wastewater treatment system as a whole and ensure a vessel's legislative compliance.

6. WASTEWATER SYSTEM SIZING AND AWTP SELECTION

Problem No.4 – the high quantity of grey water generated on yachts during full load operations.

Yachts generally use a significantly larger amount of water on a day-to-day basis than any other type of ship. Applications such as galley and laundry as well as sanitary water use for washing and showering mean that typical values for a yacht 'passenger' are in the region of anything up to 500 litres/person/day. This compared to a cruise ship of 160-200 litres and a commercial vessel of 110-150 litres.

But if the sizing is calculated on these figures alone the selection can, in many cases, be overly large with a potential for dramatic under loading during periods of down-time. So, as well as considering dual units or 2-in-1 functionality, system specifiers should consider hydraulic loading calculations for crew and VIPs separately if it is accepted that the crew will use less water per person than VIPs.

A full understanding of the yachts anticipated operating profile, manning patterns and operational requirements can mean that more efficient and practical design solutions can be proposed. This is the 'whole ship' concept adopted as standard for every ACO Marine project where a full understanding of the vessel and its wastewater sources means the best fit for the vessel - not just in technical and space terms, but financially also.

A case in point was a retrofit project on one of the world's largest megayachts. It currently has out-dated and non-compliant units installed and is therefore very limited as to where it can operate. As has been already mentioned, there is never enough space and retrofits usually require the replacement of older conventional systems that treat only the black water to be replaced not only by 227(64) [1] 4.2 compliant units but ones that will also treat the ship's hitherto untreated grey water. This is the classic "*quart into a pint pot*" scenario and in the case of the subject yacht there simply was not the space to provide for full treatment of full passenger load of all wastewater.

The concept of "size smart, size right" means that everything on the vessel has to be taken into consideration. The result of the analysis in this particular case was that two units of quarter the full size requirement (i.e. total 50% treatment capacity at full manning levels) could be installed following some judicial redesign of ship tankage creating significantly more holding capacity. This meant that, during full load periods, surplus wastewater that could not be treated by the two MBR units was held in the collecting tanks. When the passengers disembarked, both units were kept on line until the collecting tanks were emptied. Then one unit could be shut down with the remaining unit being more than sufficient to meet the demands of the 'crew only' generated wastewater.

With this solution calculations were developed to detail exactly how long the yacht could be at full manning load before the tanks would be filled and this was, based on the known operating profile of this particular yacht, more than adequate and so the reduced MBR size selection was accepted and approved.

In this case the project did not go ahead as the whole refit has been postponed, but the principle still applies and shows how flexibility in design and approach can reap significant dividends.

7. CONCLUSIONS

Wastewater management on Super and Mega Yachts is a challenging dilemma often not addressed early enough in the design stages to the point where, when eventually considered, compromises are forced on designers and manufacturers alike to propose and install a solution that has a higher risk of failure. This in turn may then directly impact the vessels operational flexibility.

Wastewater Treatment plants, both STP and AWTP, are relatively large pieces of equipment preferably located in the bottom of the ship; getting it wrong can be a very costly mistake, often very difficult or expensive to resolve. Naval Architects, Yacht Designers and Yacht Builders should avail themselves of the services of reputable wastewater solution providers at the earliest opportunity. As has been demonstrated in this paper, ACO Marine will work together as a reliable partner to all those involved in the yacht construction, providing 3D component, and even compartment, modelling to ensure the optimum fit for the optimum solution whilst at the same time offering the full 'cradle to grave' support philosophy ensuring that wherever they are in the world, Yacht staff can always get the help and support they need to ensure trouble-free wastewater management on their vessels.

MEPC 227(64) Section 4.2 compliant MBR systems offer a truly 'future-proofed' solution to all current and anticipated legislation. The days for other, currently more popular, alternative chemical and dilution systems are seriously numbered. No yacht should find itself in a situation where it is prohibited from either sailing from, or into, any given port anywhere in the world. Notwithstanding any extreme 'local' rules, use of correctly designed, installed and maintained MBR AWTP units should ensure trouble-free sailing for the life of the yacht.

8. REFERENCES

1. International Maritime Organisation IMO, 2012 Guidelines on Implementation of Effluent Standards and Performance Tests for Sewage Treatment Plants, Resolution MEPC 227(64), 5th October 2013

2. International Maritime Organisation IMO, Amendments to the Annex of the Protocol of 1978 Relating to the International Convention for the Prevention of Pollution from Ships, 1973 (Special Area Provisions and the Designation of the Baltic Sea as a Special Area Under MARPOL Annex IV), 15th July 2011

3. International Maritime Organisation IMO, Revised Guidelines on Implementation of Effluent Standards and Performance Tests for Sewage Treatment Plants, 13th October 2006

9. AUTHORS BIOGRAPHY

Mark Beavis holds the position of Managing Director of ACO Marine s.r.o. in the Czech Republic and also its daughter company, ACO Marine Systems GmbH in Germany. He is responsible for all aspects of product development, innovation and market development of their in-house designed and manufactured complete wastewater management packages.

Mark studied Marine Engineering at Solent University, Southampton and served as Marine Engineering Officer on Bulk Carriers during the late 1970s and 1980s. In 1987 he served briefly as Engineering Superintendent with Kappa Maritime in London before joining the Sea Fisheries Inspectorate working with the Fishery Protection Squadron of the Royal Navy. For the last twenty years he has specialised in marine wastewater management, beginning in 1996 when he joined Hamworthy Wastewater Systems Limited where, as Marine Sales Manager, he travelled globally, developing bespoke wastewater management solutions for all vessel types including yachts, military and offshore vessels.

In 2006 he was invited to join the ACO Group to head up its newly formed marine division, ACO Marine, as Managing Director.

Mark is a Fellow of the Institute of Marine Engineering, Science and Technology (FIMarEST) and a registered Marine Engineer (IMarEng).

PASSENGER YACHTS OUTFITTING MATERIALS, CERTIFICATION AND FIRE LOAD CALCULATION

C Russo, Lloyd's Register, Italy

SUMMARY

The introduction of the Passenger Yacht Code (PYC) has increased the complexity for the Designer, Interior Architects, Owners and their representatives in the choice of materials outfitting within accommodation spaces, service spaces and control stations. Such requirements, although familiar to most of the Cruise Ship architects and Companies, are relatively new for the Yacht architects, Yacht Builders and Owners, as the level of finish on such vessel is generally far more luxurious when comparing it to a commercial vessel and sometimes the products sought are not certified. The regulations within the PYC and the research for alternative and innovative products, which however should produce a finish product agreeable to the Owners, are in some cases problematic not only for their application but also for what concerns the interpretation of the legislation.

NOMENCLATURE

DLL	Davit Launch Liferaft
FTP	International Code for Application of Fire Test Procedures
IMO	International Maritime Organization
LL	Load Lines Convention
LYC	Large Yacht Code (3rd Edition)
LSA	Life Saving Appliances
MED	Marine Equipment Directive
MES	Marine Evacuation System
PYC	Passenger Yacht Code (6th Edition)
SOLAS	Safety of Life at Sea
STCW	Standard of Training, Certification and Watch keeping
TA	Type Approval Certificate

1. INTRODUCTION

Since its inception in 2010, the PYC [1] has sought to rationalise the requirements and standards to be met by a pleasure yacht of any size which carries more than 12 but not more than 36 passengers on international voyages, particularly with respect to SOLAS [2] , LL 1966 and STCW. The PYC sets out technical, safety and operational standards appropriate to the size and operation of the so-called 'Passenger Yachts' and the standards incorporated are largely based on international conventions applying to commercial vessels with the inclusion of equivalencies where it is not reasonable or practicable to comply with the conventions. The philosophy of the Code and its structure provides a "stepped approach" to be adopted in ascertaining appropriate standards to be applied to the various categories of yacht and their operational areas, and although could be considered as a 'natural progression' from LYC [3], it is 'philosophically' very different. Another important aspect of the PYC is to consider the vessel being its own best survival craft, thus in general terms, where the carriage of DLLs or MES are permitted in lieu of lifeboats, enhanced standards of subdivision, damage stability and fire protection are applied. Within the fire safety objective in mind the PYC introduce requirements for material used for interior construction components to ensure that the fire safety functional requirement, such as but not limited to the ones below, are achieved:

- Restricted used of combustible materials;
- Containment of any fire in the space of origin;
- Protection of means of escape

Such requirements are either achieved by ensuring compliance with the prescriptive requirements specified in PYC Chapter 6, or by alternative design and arrangement which comply with the guidelines developed by IMO. The prescriptive requirements within PYC Chapter 6 for outfitting materials are largely based on SOLAS Passenger Ship and FTP Code Requirements and are more onerous than those prescribed by the LYC, which could (in many circumstances), be alleviated by the provision of a fully addressable fire detection system and automatic sprinkler in accommodation and service areas. Also the PYC is more restricted to SOLAS (applicable to passenger ship for less than 36 passengers) as it is required that a fixed fire-extinguishing system and a fixed fire detection and fire alarm system is provided in all control stations, accommodation and service spaces. Such requirement is to be considered as a 'trade-off' for what concerns the carriage of lifeboats as aforementioned and also in order to permit some relaxations in the overall fire, smoke and toxicity characteristic of the materials being used in luxurious areas for the Owner and Guests.

2. USE OF MATERIALS IN THE ACCOMMODATION AND SERVICE SPACE

In order to achieve the functional requirements of PYC Chapter 6, one of functional requirements is the restricted use of combustible materials on-board. Such objective is achieved mainly by providing each single space on-board with non-combustible structures (e.g. lining, ceiling, grounds etc....). The objective is to permit the safe escape of the occupants which would be otherwise impeded by the collapsing of such structure. However on such surfaces, a minimum application of combustible material is permitted, depending on the space and its purposes, as for instance there are provisions in spaces such as sauna whereby it is permitted the use of traditional wooden lining and ceilings. On Passenger Yachts, “A”, “B” or “C” class divisions in accommodation spaces and service spaces and cabin balconies which are faced with combustible materials, facings, moulding, decorations and veneers shall have a calorific value not exceeding 45MJ/m2 of the area for the thickness used and the total volume of combustible facings, moulding, decorations and veneers shall not exceed a volume equivalent to 2.5mm veneer on the combined are of the walls and ceiling linings. Moreover the exposed surfaces in corridor, stairway enclosures and of bulkhead and ceiling linings in accommodation, service spaces and control stations, surface and grounds in concealed or inaccessible space in accommodation, control station and service spaces shall have low flame-spread characteristic. Further to the above there are requirements regarding the quantity of smoke and toxic products released from combustible materials, including surface finishing. The purpose of this regulation is to reduce the hazard to life from smoke and toxic products generated during a fire in spaces where persons normally work or live. In this respect, paints, vanishes and other finished used on exposed interior surface shall not be capable of producing excessive quantities of smoke and toxic products as determined in accordance to the FTP Code (Annex 1 Part 2). Also primary deck coverings applied in accommodation and service spaces and control station shall be of approved material which will not give rise to smoke or toxic explosive hazards at elevated temperature in accordance to the FTP Code (Annex 1 Part 5)

3. FTP CODE

The FTP Code has been developed in order to collect and summarise in a single document all test procedures in testing products such as bulkheads, decks, fire doors, windows, linings, penetrations and the test procedures identify the test methods and the acceptance and classification criteria.

3.1 FTP CODE ANNEX 1, PART 1 NON-COMBUSTIBILITY TEST

The non-combustibility is verified in accordance with the ISO Standard 1182 (which is normally applied also to other industries such as building and transport) and the FTP Code establish the acceptance criteria which are based on the principle that the product does not contribute to the development of the fire even though there is a limited margin of combustion as the product is exposed to a temperature of 750°C. A product which has passed such test is deemed also in compliance with the required low-flame spread characteristic and therefore can be installed on-board without additional tests. In general, products made only of glass, concrete, ceramic products, natural stone, masonry units, common metals and metal alloys are considered as being non-combustible and they may be installed without testing and approval. However, special attention is to be made to the fact that such materials do not contain combustible additives which may impair the original characteristics.

3.2 FTP CODE ANNEX 1, PART 2, SMOKE AND TOXICITY TEST

The test is generally divided in two parts; part 1 has the purpose to measure the smoke production from the exposed surface of specimens of essentially flat materials, composites or assemblies not exceeding 25 mm in thickness, when placed in a horizontal orientation and subjected to specified levels of thermal irradiance in a closed cabinet with or without the application of a pilot flame whilst part 2 measures the concentrations of toxic gases generated using Fourier transform infrared spectroscopy (FTIR). In such test, it important to pay attention to the sample tested and to its composition, especially when various layers and/or materials intended to be applied may affect the test results. As already mentioned, a material in compliance with the FTP Code Annex 1 Part 1 is considered to be compliant to this part without further testing, whilst in general, surface materials and primary deck coverings with both the total heat release Q_t of not more than 0.2 MJ and the peak heat release rate (Q_p) of not more than 1 kW (both values determined in accordance with part 5 of annex 1) are considered to comply with the requirements of this part without further testing and expected to satisfy the requirement of maximum gross calorific value (e.g., 45 MJ/m^2) without further testing.

3.3 FTP CODE ANNEX 1, PART 5, TEST FOR SURFACE FLAMMABILITY (TEST FOR SURFACE MATERIALS AND PRIMARY DECK COVERINGS)

Materials having the characteristic of low flame spread are not deemed non-combustible and therefore to be considered when calculating the allowable combustible volume in the space. Such materials have the properties of being resistant to the engraftment of the flame and to its spreading in surface. Such property has a certain importance during the initial phase of a fire while is virtually nil in the development phase of a fire. Depending on the material application, various acceptance criteria are established. The test provides methods for evaluating flammability characteristics of 155 mm x 800 mm specimens in vertical orientation and such specimens are exposed to a graded radiant flux field supplied by a gas-fired radiant panel. Means are provided for observing the times to ignition, spread and extinguishment of flame along the length of the specimen as well as for measuring the compensated millivolt signal of the stack gas thermocouples as the burning progresses. Experimental results are reported in terms of: heat for ignition, heat for sustained burning, critical flux at extinguishment and heat release of specimen during burning. Materials having average values for all of the surface flammability criteria that comply with the values as listed in *figure 1* are considered to meet the requirement for low flame-spread in compliance with the relevant regulations.

Table 1 – Surface flammability criteria

	Bulkhead, wall and ceiling linings	Floor coverings	Primary deck coverings
CFE (kW/m²)	≥20.0	≥7.0	≥7.0
Qsb (MJ/m²)	≥1.5	≥0.25	≥0.25
Qt (MJ)	≤0.7	≤2.0	≤2.0
Qp (kW)	≤4.0	≤10.0	≤10.0
Burning droplets	Not produced	No more than 10 burning drops	Not produced

Where:

CFE= Critical flux at extinguishment
Qsb= Heat for sustained burning
Qt= Total heat release
Qp= Peak heat release rate

Note: Qsb means an average of heat for sustained burning, as defined in paragraph 9.3 of appendix 1.

Figure 1

Particular attention should be paid to the substrate of surface material and floor covering which shall be tested using their full thickness, attaching them, by means of an adhesive if appropriate, to the substrate to which they will be attached in practice as the test specimen shall reflect actual application and for the substrate of primary deck covering which shall be applied to a steel plate having a thickness of 3 ± 0.3 mm. The specimens shall have a nominal thickness; the components and construction of the primary deck covering shall reflect actual application. It is to be specified that a "primary deck covering" is the first layer of a floor construction which is applied directly on top of the deck plating and is inclusive of any primary coat, anti-corrosive compound or adhesive which is necessary to provide protection or adhesion to the deck plating. Other layers in the floor construction above the deck plating are "floor coverings". When the product that is the first layer of a floor construction which is applied directly on top of the deck plating and is also the exposed surface (i.e. no other layer applied on it), it shall be considered as the "floor covering", and shall comply with the requirements of "floor covering". If the floor covering has a multilayer construction, the Administration may require the tests to be conducted for each layer or for combinations of some layers of the floor coverings. Each layer separately, or a combination of layers (i.e. the test and approval are applicable only to this combination), of the floor covering shall comply with this part. Surface materials and primary deck coverings with both the total heat release *(Qt)* of not more than 0.2 MJ and the peak heat release rate *(Qp)* of not more than 1.0 kW are considered to comply with the requirements for smoke and toxicity without further testing.

3.4 FTP CODE ANNEX 1, PART 7, TEST FOR VERTICALLY SUPPORTED TEXTILES AND FILMS

This test qualifies textiles and films used primarily as vertically hanging curtains and draperies, as meeting the requirements for the resistance to propagation of flame specified in the relevant regulations of the Conventions. Fabrics, which are not inherently flame resistant, shall be exposed to cleaning or exposure procedures and tested both before and after such treatment. The test method provides information on the ability of a fabric to resist sustained ignition and flame propagation when exposed to a small igniting flame. The test is carried out with a gas burner so mounted that the axis of the burner barrel is capable of adjustment to each of three fixed positions, viz., vertically upwards, horizontal or at an angle of 60° to the horizontal.

3.5 FTP CODE ANNEX 1, PART 8, TEST FOR UPHOLSTERED FURNITURE

This test procedure prescribes methods for assessing the ignitability of material combinations, e.g., covers and filling used in upholstered seating when subjected to either a smouldering cigarette or a lighted match as might be applied accidentally in the use of upholstered seats. It does not cover ignition caused by deliberate acts of vandalism. The principle is to subject an assembly of upholstery materials arranged to represent, in stylized form, the joint between the seat and back (or seat and arm) surfaces of a chair to two sources of ignition, one being a smouldering cigarette, and the other a flaming source approximating to the calorific output of a burning match. Materials installed in accommodation and public spaces (not in room required to be containing furniture and furnishings of restricted fire risk because for instance forming part of escape routes) need not to comply with the requirements of this test.

3.6 FTP CODE ANNEX 1, PART 9, TEST FOR BEDDING COMPONENTS

Similarly to upholstered furniture, bedding components installed in accommodation and public spaces (not in room required to be containing furniture and furnishings of restricted fire risk because for instance forming part of escape routes) need not to comply with the requirements of this test. The bedding component is classified as not readily ignitable if it shows no progressive smouldering ignition or flaming ignition within some parameters. The test specified in this method deals with a simple presentation of a particular aspect of the potential fire situation typified by smouldering cigarettes and a match equivalent flame exposure to the product similar to the test for upholstered furniture.

4. FURNITURE

A separate chapter is to be dedicated to furniture which are generally defined as all on-board decorating accessories, fixed and non, such as: chairs, armchairs, sofas, benches, tables, desks, counters, bedside tables, chests of drawers, bed frame, libraries, wardrobe, tv standings fire risk, book shelving etc.... which are generally independent from the wall/ceiling structure, that is that none of their part is communal with the wall/ceiling and/or participates to the closure of such structures. In such circumstances there are not many restrictions or prescription to the use of furniture in accommodation and public spaces such as cabins, lounges, dining rooms etc.... unless such spaces are part of an escape route (*PYC 6.7.1 (d) refers*) or adjacent to the LSA *(PYC 6.7 (23) refers)* unless their risks has been sufficiently evaluated in accordance with IMO MSC.1/Circular 1274 which provides Guidelines for the Evaluation of fire risk of External Areas on Passenger Ships. A special mention is to be addressed to those furniture on cabin balconies, which following the fire on 'Star Princess' back on 2006, are required to be properly addressed. In this case, the PYC is less restrict than SOLAS > 36 passenger as, for those balconies not adjacent to the LSA, and for which a risk assessment has been carried out, may permit the use of non-restricted fire risk furniture on cabin balconies provided it is possible to readily direct a jet of water, for firefighting purposes, from the deck immediately above or adjacent. Restricted furniture and furnishing of restricted fire risk, for the purpose of PYC are:

- case furniture such as desks, wardrobes, dressing tables, bureaux, or dressers are constructed entirely of approved non-combustible materials, except that a combustible veneer not exceeding 2 mm may be used on the working surface of such articles;
- free-standing furniture such as chairs, sofas, or tables, are constructed with frames of non-combustible materials;
- draperies, curtains and other suspended textile materials have qualities of resistance to the propagation of flame not inferior to those of wool having a mass of 0.8 kg/m^2 , this being determined in accordance with the FTP Code
- floor coverings have low flame-spread characteristics;
- exposed surfaces of bulkheads, linings and ceilings have low flame-spread characteristics;
- upholstered furniture has qualities of resistance to the ignition and propagation of flame, this being determined in accordance with the FTP Code;
- bedding components have qualities of resistance to the ignition and propagation of flame, this being determined in accordance with the Fire Test Procedures Code

The above guidelines for accommodation space do not applies to stairway enclosure and corridor, in virtue of the fact that such spaces are to be remained as the most safe spaces on board, as to allow all passenger and crew to reach the embarkation station. Hence, furniture in stairway enclosures shall be limited to seating. It shall be fixed, limited to six seats on each deck in each stairway enclosure, be of restricted fire risk determined in accordance with the FTP Code, and shall not restrict the passenger escape route. The Administration may permit additional seating in the main reception area within a stairway enclosure if it is fixed, non-combustible and does not restrict the passenger escape route. In corridors, furniture shall not be permitted in corridors forming escape routes in cabin areas. The only expectations, are for stairways the provision of lockers of non-combustible material, providing storage for non-hazardous safety equipment and, in corridors, drinking water dispensers and ice cube machines may be permitted provided they are fixed and do not restrict the width of the escape routes. This applies as well to decorative flower or plant arrangements, statues or other objects of art such as paintings and tapestries in corridors and stairways. Artworks are works such as paintings, statues, artistic vases, tapestries, mosaics, antiques and any other decorative element that is not an integral part of walls, ceilings or furniture, and which are generally 'unique' in their applications and not so large to cover a big portion of the structure in subject.

5. OTHERS

There are other materials and or decorative elements within the Codes which are subject to the Codes and Conventions herewith describes, such as waste receptacles, window boxes, mouldings, skirting boards, light fixtures etc....., and for each, depending on which space they are installed, there are prescriptions and/or allowance, which are summarised in the Annex 4 of the PYC Code in the form of a table as represented in *figure 2.*

REG 13-36 PASSENGER YACHT CODE

ANNEX 4
GUIDANCE MATERIALS MATRIX FOR ACCOMMODATION SPACES, SERVICE SPACES AND CONTROL STATIONS

Type of Space	Control Stations	Stairway Enclosures	Corridors	Accommodation Areas	Service spaces and Garage Spaces	Balconies see note (10)	Included in Calorific value Calculation	Included in Fire Volume Calculation
Category Of Space / Material	1	4	2	3	6, 9, 11	10	1 - 9	1 - 9
(A) Floor coverings	Annex 1, Part 2	Annex 1, Part 5 Annex 1, Part 2	Annex 1, Part 5 Annex 1, Part 2	Annex 1, Part 2 (PYC 6.4 refers)	Annex 1, Part 2	Annex 1, Part 5 Annex 1, Part 2 Except natural hardwood decking systems See Note (10)		
(B) Primary deck coverings	Annex 1, Part 5 Annex 1, Part 2	Annex 1, Part 5 Annex 1, Part 2	Annex 1, Part 5 Annex 1, Part 2	Annex 1, Part 5 Annex 1, Part 2	Annex 1, Part 5 Annex 1, Part 2	Annex 1, Part 5 Annex 1, Part 2		
(C) Exposed surfaces of walls and ceilings, panel surfaces and ceiling linings, including paints, varnishes, fabrics & veneers (see note 16)	Annex 1, Part 5 Annex 1, Part 2 see notes (5)	Annex 1, Part 5 Annex 1, Part 2 see notes (5)	Annex 1, Part 5 Annex 1, Part 2 see notes (5)	Annex 1, Part 5 Annex 1, Part 2 see notes (5)	Annex 1, Part 5 Annex 1, Part 2 see notes (5)	Annex 1, Part 5 Annex 1, Part 2 see notes (5)	✓	✓ (except Balconies)
(D) Skirting Board	Annex 1 Part 2 Annex 1 Part 5 see notes (5) & (6)	Annex 1 Part 2 Annex 1 Part 5 see note (6)	Annex 1 Part 2 Annex 1 Part 5 see note (6)	Annex 1 Part 2 Annex 1 Part 5 see note (6)	Annex 1 Part 2 Annex 1 Part 5 see note (6)	No Applicable Requirements		✓
(E) Combustible facings, mouldings, decorations	Annex 1 Part 2 Annex 1 Part 5 see notes (5) & (6)	Annex 1 Part 2 Annex 1 Part 5 see note (6)	Annex 1 Part 2 Annex 1 Part 5 see note (6)	Annex 1 Part 2 Annex 1 Part 5 see note (6)	Annex 1 Part 2 Annex 1 Part 5 see note (6)	Annex 1 Part 2 Annex 1 Part 5 see note (6)		✓
(F) Partial Bulkhead	Annex 1, Part 1	Annex 1, Part 1	Annex 1, Part 1	Annex 1, Part 1	Annex 1, Part 1	Annex 1, Part 1		
(G) Wallpanel/Lining	Annex 1, Part 1 see note (15)	Annex 1, Part 1 see note (15)	Annex 1, Part 1 see note (15)	Annex 1, Part 1 see note (15)	Annex 1, Part 1 see note (15)	Annex 1, Part 1 see note (15)		
(H) Concealed panel surface (see note 17)	Annex 1, Part 5	Annex 1, Part 5	Annex 1, Part 5	Annex 1, Part 5	Annex 1, Part 5	No Applicable Requirements	✓	
(I) Draught stops	Annex 1, Part 1 see note (15)	Annex 1, Part 1 see note (15)	Annex 1, Part 1 see note (15)	Annex 1, Part 1 see note (15)	Annex 1, Part 1 see note (15)	Not applicable		
(J) Grounds & supports (see note 18)	Annex 1, Part 1 see notes (14) & (15)	Annex 1, Part 1 see notes (14) & (15)	Annex 1, Part 1 see notes (14) & (15)	Annex 1, Part 1 see notes (14) & (15)	Annex 1, Part 1 see notes (14) & (15)	Annex 1, Part 1 see notes (14) & (15)		
(K) Insulation	Annex 1, Part 1 see note (12)	Annex 1, Part 1 see note (12)	Annex 1, Part 1 see note (12)	Annex 1, Part 1 see note (12)	Annex 1, Part 1 see note (12)	Annex 1, Part 1 see note (12)		
(L) Insulation surface	Annex 1, Part 5 see note (13)	Annex 1, Part 5 see note (13)	Annex 1, Part 5 see note (13)	Annex 1, Part 5 see note (13)	Annex 1, Part 5 see note (13)	Annex 1, Part 5 see note (13)		
(M) Curtains and other suspended textiles	Annex 1, Part 7	Annex 1, Part 7	Annex 1, Part 7	Annex 1, Part 7	Annex 1, Part 7	Not applicable		
(N) Upholstered furniture (including Sofas & Chairs)	see note (5)	see notes (2) & (4)	See Note (11)	No Applicable Requirements	No Applicable Requirements	Annex 1, Part 8 see note (9)		
(O) Case furniture (desks, wardrobes, dressing tables, bureaux, dressers etc)	see note (5)	Not allowed	Not allowed	see note (17)	see note (17)	No Applicable Requirements		
(P) Free standing furniture (tables etc)	see note (5)	Not allowed	Not allowed	see note (17)	see note (17)	see note (9)		
(Q) Window box surface	Annex 1, Part 5 Annex 1, Part 2 see note (1)	Annex 1, Part 5 Annex 1, Part 2 see note (1)	Annex 1, Part 5 Annex 1, Part 2 see note (1)	Annex 1, Part 5 Annex 1, Part 2 see note (1)	Annex 1, Part 5 Annex 1, Part 2 see note (1)	Annex 1, Part 5 Annex 1, Part 2 see note (1)	✓	✓ see note (1)
(R) GRP surfaces in private Bathroom/Toilet Modules	Not applicable	Not applicable	Not applicable	Annex 1, Part 5 Annex 1, Part 2	Not applicable	Not applicable	✓	✓
(S) Light fixtures	In general no certification required, but to be looked at on a case by case basis, in particular when attached to ceiling or fixed to bulkhead							
(T) Waste receptacles	Not allowed	Not allowed	Not allowed	Non-combustible	Non-combustible see note (3)	Non-combustible		
(U) Object d'art, plasters, paintwork (NEVER TO REDUCE ESCAPE WIDTHS)	Not allowed	see notes (7) & (8)	see note (7)	No Applicable Requirements	Not allowed	No Applicable Requirements		

(to be read in conjunction with Chapter 6 of the Code & MSC/Circ. 1120, Annex, Interpretation of SOLAS Regulations II-2/5.3 & 6. For guidance only, actual requirements should be taken from the text within chapter 6 of the Code)

(Page 250 of 254)

(Sixth Edition) (January 2016)

Figure 2

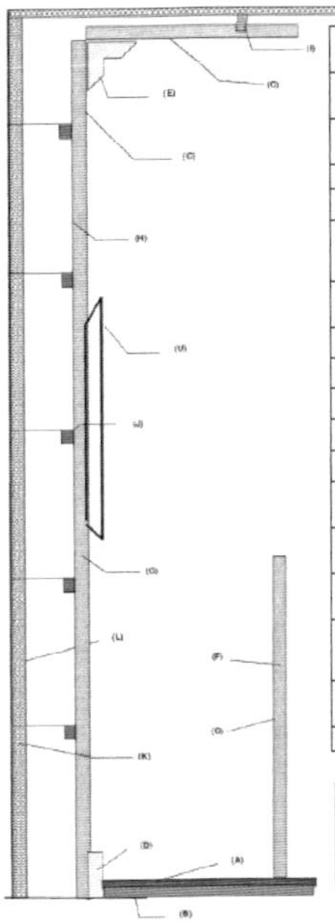

Note that where the words 'No Applicable Requirements' appear, although no compliance is required by the Regulations, use of materials having restricted fire risk as per applicable requirements, is recommended	
note(1):	The windows box has to have a non-combustible structure. Where the strucutre is faced with a combustible veneer, this item is to be included in the fire volume calculation.
note(2):	Furniture in stairway enclosures shall be limited to seating. It shall be fixed, limited to six seats on each deck in each stairway enclosure, of restricted fire risk, and shall not restrict the passenger escape route.
note(3):	Reference can be made to IACS UI SC166
note(4):	These items are to be constructed with frames of non-combustible materials and, if upholstered are to comply with FTP Annex 1, Part 8
note(5):	Such spaces were not originally meant to be outfitted as accommodation spaces and fire volume calculations were in fact not required. Should they be constructed as such they should be treated as spaces of restricted fire risk as defined in SOLAS II-2/3.40
note(6):	FTP Code, Annex 1, Parts 2 and 5 are applicable only to paints, varnishes and other finishes (Reg. II-2/6.2) (i.e. not veneers or substrate).
note(7):	To be kept to a minimum, be fixed and have properties that are not readily ignitible.
note(8):	Large paintings that form the majority of the bulkhead are to have low frame spread surface.
note(9):	Refer to PYC 6.3 (21),(22) & (23) for details regarding Furniture and furnishings on cabin balconies.
note (10):	For open deck spaces other than balconies attention is to be paid to MSC.1/Circ.1274
note (11):	Not allowed in cabin corridors. In other corridors (i.e. corridors not forming part of an Escape Route), notes 2 and 4 apply.
note (12):	Insulating materials located in cargo spaces, mail rooms and refrigerated compartments of service spaces need not be non-combustible. SOLAS II-2/5.3.1.1 refers.
note (13):	Combustible vapour barriers and adhesives used in conjunction with insulation shall be kept to the minimum quantity practicable. SOLAS II-2/5.3.1.1 refers.
note (14):	Linings, grounds, draught stops and ceilings in mail rooms, baggage rooms, saunas or refrigerated compartments need not be non-combustible. SOLAS II-2/5.3.1.2.1 refers.
note (15):	Grounds in concealed or inaccessible spaces in accommodation, service spaces and control stations shall have low flame spread characteristics in accordance with FTP Code Annex 1, Part 5. SOLAS II-2/5.3.2.4.1.2 refers.
note (16):	See PYC 6.3 (18) for details regarding the possible relaxation of Low Flame-spread characteristics in areas appropriated for the use of the Owners anf Guests, subject to agreement by the Flag Administration.
note (17):	Refer to PYC 6.7(1) (i) & (ii) for detailed requirements

Lloyd's Register

The content of Annex 4 is generously provided by Lloyds Register to the appreciation of the Red Ensign Group (REG).

6. ALLOWABLE FIRE VOLUME CALCULATION [4]

To demonstrate compliance with the applicable Codes and Conventions above mentioned and relative material which may be used in accommodation spaces, service spaces and control stations on passenger yacht, fire load calculations are to be submitted to the recognised organisations and to the competent maritime administration. The calculations are to be submitted together with the architectural drawings, material certificates and specifications and relative sample book. The calculation is to be carried out for all public area and for each cabin type. Generally such calculations are not required for spaces such as pantries (not containing cooking appliances), laundries, magazines, workshops, crew stairways, lift, sauna etc.... unless such spaces are built to a 'luxury' standards (e.g. with materials which are not of restricted fire risks etc.... and which have a 'passenger' finish). There are two fire load calculation methods available, the fire volume method (SOLAS II-2 Reg. 5.3) or the mass method (MSC/Circ. 1003). Either one of these methods can be used in way of a space to show compliance but it needs to be clear when the calculation is submitted which method has been used for the space concerned. Currently the MSC/Circ. 1003 mass method is not generally used to show compliance for fire load calculation but it is an option which is taken in consideration for particular space, arrangements. In the so-called volume method, it is calculated the total volume of the maximum allowable combustible material which is determined by adding all surface of walls and ceiling and multiplying such value for 0.0025. Combustible materials are those which constitute the exposed surface of walls and ceilings, plus moulding, skirting boards and similar. These materials are those certified (i.e. low-lame spread) and non-certified. Such method is generally submitted in a template as per *figure 3*:

ALLOWABLE FIRE VOLUME CALCULATION — Page 01

Hull: #001

Area Main Contractor: XYZ — Publ. Space: SMALL LOUNGE — Deck: 6 — Frames: 32 42

A			B	C	D	E	F	G
Material description and code	**Code**	**Certificate**	**Area / Surface** (m2)	**Thickness / Length** (m)	**Volume** (m3)	**Density** (kg/m3)	**Calorific Vol.** (MJ/kg)	**Cal. Value** (MJ/m2))
Bulkheads:								
e.g.Resopal Laminate	125	MED 1254584	42.0	0.0010	0.042	1400	18	25
Ceilings								
e.g. Abet Laminate	254	MED 1526842	120.0	0.0009	0.108	2500	18.26	41
e.g. Painted metal blue	504	MED1945782	80.00	0.0040	0.320	1500	4.5	27
				Tot. VOLUME:	0.470			
Wall surface			125.0					
Ceiling surface			200.0					
Tot. SURFACE:			325.0					
				Acceptable FIRE VOLUME (Tot. SURF. X 0,0025):	0.813			

Whereby:

Column	Type	Description
A	Position – Material	Material Code and its application.
B	Area – Surface	Square meters utilised.
C	Thickness / (length)	Thickness of the material (in accordance with Certification)
D	Volume	Volume obtained by **B*C**
E	Density	Density (as per certification)
F	Calorific Volume	Calorific volume expressed in MJ/Kg (as per certification)
G	Calorific Value	Calorific Value obtained by **C*E*F**, which is to be less than 45 MJ/sq.

Figure 3

The calculation should show that the total volume of the materials in column **D** is lesser than the total volume admissible for the space. If greater, actions should be taken to reduce the amount of combustible material within the space, or to introduce alternative solutions, such as 'fire-breaks', which are to be agreed with the intended Administration.

In the mass method, instead, the total mass of combustible materials of each enclosed space is calculated based on the following formula:

$$TCMPA = M_{CMBST} / A$$

where:

TCMPA is the total mass of combustible materials per unit area of the space (kg/m^2);

M_{CMBST} is the total mass of combustible materials of the space (kg); and

A is the floor area of the space (m^2).

In such method all construction materials such as cable insulation, plastic pipes, veneers and combustible materials permitted to be used under the relevant regulations; and outfitting which may be installed during the construction or provided by the ship's owner or crew, including furniture, bedding components and electrical appliances, should be included. Maximum values for the total mass of combustible materials per unit area TCMPA are provided in the *figure 4* below

Type of space	Passenger ships carrying more than 36 passengers	Passenger ships carrying not more than 36 passengers	Cargo ships
Stairways, corridors	5 kg/m^2	5 kg/m^2	5 kg/m^2
Control stations	5 kg/m^2	5 kg/m^2	5 kg/m^2
Accommodation spaces	Minor fire risk15 kg/m^2	35 kg/m^2	35 kg/m^2
	Moderate fire risk35 kg/m^2		
	Greater fire risk36 kg/m^2		
Service spaces surrounded by "A" class divisions	45 kg/m^2	45 kg/m^2	46 kg/m^2

Figure 4

7. DEFINITIONS

Accommodation spaces are those spaces used for public spaces, corridors, lavatories, cabins, offices, hospitals, cinemas, game and hobby rooms, barber shops, pantries containing no cooking appliances and similar spaces;
Non-combustible material is a material which neither burns nor gives off flammable vapours in sufficient quantity for self-ignition when heated to approximately 750°C, this being determined in accordance with the Fire Test Procedures Code;
Combustible material is any material other than a non-combustible material;
Public spaces are those portions of the accommodation which are used for halls, dining rooms, lounges and similar permanently enclosed spaces;
Service spaces are those spaces used for galleys, pantries containing cooking appliances, lockers, mail and specie rooms, storerooms, workshops other than those forming part of the machinery spaces, and similar spaces and trunks to such spaces;
Cabin balcony is an open deck space which is provided for the exclusive use of the occupants of a single cabin and has direct access from such a cabin.

8. CONCLUSIONS

As passenger yachts are getting more and more complex, the choice of materials for outfitting is a critical one, as beside to compliance with the applicable Codes and Conventions, materials and finishing should be to a standard acceptable to the Owner. As such, all stakeholders, from interiors designer to the intended Recognised Organisation and Administration should work together, well in advance, to determine the characteristic and properties of the intended materials and their relative installation and whether or not they are suitable for their purpose. Once the choice is made, it should be ensured that the material is duly certified and applied in accordance with its MED and/or TA certification and that calculation to show compliance are prepared. Such calculation, together with the material certification should be included in the so-called 'Sample book' which should contains a sample of each material utilised, used space by space, as to permit the verification by the attending Surveyor. In the future this will also be a game changer for those shipyards specialised in re-fittings, as such sample book and fire volume calculation will be need to be prepared for major refits where the outfitting of internal spaces in passenger yachts are envisaged, a practice not normally common nowadays for yacht built under LYC.

9. ACKNOWLEDGEMENTS

The author would like to thank all current and formerly employees of Lloyd's Register, particularly fire and safety experts for their contribution, both direct and indirect, in assisting creating this paper through their mentoring and assistance in the past two years. My thanks also extend to all shipyards and contractors, designers, interior architects etc…., with whom I have worked and share experience and knowledge in the course of my career.

10. REFERENCES

References should be numbered in the text where they occur, with the number in square brackets [1] and a complete list given at the end of the paper.

Cross Referencing, footnotes, endnotes and 'Smart Tags' must NOT be used for figures and/or references.

1. The Red Ensign Group – A Code of Practice for Yachts Carrying 13 to 36 passengers (The Passenger Yacht Code) – Sixth Edition – January 2016;

2. International Maritime Organisation – International Convention for the Safety of Life at Sea (SOLAS) 1974, as amended;

3. Maritime and Coastguard Agency – The Large Commercial Yacht Code (LY3) – 3rd Edition – August 2012;

4. LR Summary of information relating to fire load calculations for ship builder guidance including passenger yacht code references – John Elmer, Stefano Russo– April 2014.

11. AUTHORS BIOGRAPHY

Carlo Russo holds the current position of Senior Specialist, Fire & Safety at Lloyd's Register EMEA. He is responsible for the approval of statutory aspects (fire & safety) for a variety of yachts, passenger yachts and passenger ships, both new construction and existing vessels. His previous experience includes decennial experience as certified ship's officers on tanker, gas and chemical carriers and passenger ships, and five years in UK working for Lloyd's Register Passenger Ship Support Centre for New Constructions.

IMO SECOND GENERATION INTACT STABILITY CRITERIA: APPLICATION TO A SET OF MEGA-YACHT PARENT HULLS IN A SHIP DESIGN PERSPECTIVE

N. Petacco and **P. Gualeni,** University of Genoa, IT

SUMMARY

At the IMO, the so called Second Generation Intact Stability Criteria are currently under finalization and validation. This paper describes briefly the five different stability failure modes taken into account by the rules and the multi-layered approach which defines each vulnerability level. Furthermore, all first vulnerability levels are applied to a Mega-Yacht unit and for parametric roll and pure loss of stability phenomena, also the second levels are applied. In order to understand how a design modification in terms of ship breadth may influence the susceptibility to stability failures, a set of different models beside the original one are also investigated. Investigations are carried out by means of an "in house" computational tool developed at the University of Genoa. Results are mainly analysed in terms of KG_{max} curves, in order to detect the ship feasibility domain.

NOMENCLATURE

DS	Dead ship
EA	Excessive acceleration
IMO	International Maritime Organization
ISSW	International Ship Stability Workshop
PR	Parametric roll
PLS	Pure loss of stability
SDC	Ship Design and Construction Sub-Committee
SGISC	Second Generation Intact Stability Criteria
SLF	Stability, Load Lines and Fishing Vessels Sub-Committee
SR	Surf-Riding / Broaching
STAB	International Conference on the Stability of Ships and Ocean Vehicles
A_{bk}	Overall projected area of the bilge keels (m^2)
B	Moulded breadth of the hull (m)
BMT	Transversal metacentric radius (m)
C_b	Block coefficient (-)
C_m	Midship section coefficient (-)
C_x	Prismatic coefficient (.)
Δ	Displacement (t)
d	Draft amidships (m)
g	Gravitational acceleration of 9.81 (m/s^2)
GM	Transversal metacentric height (m)
GM_c	Transversal metacentric height in still water (m)
GM_{min}	Minimum transversal metacentric height in waves (m)
ΔGM_1	Amplitude of variation of GM due to the passage of a wave along the hull (m)
GZ	Righting arm (m)
I_x	Moment of inertia referred to x axis of waterplane area (m^4)
KG	Centre of Gravity vertical position (m)
KB	Centre of Buoyancy vertical position (m)
L	Length between perpendiculars (m)
L_{OA}	Length overall (m)
M_R	Restoring moment (t m)
T_{roll}	Natura roll period of the ship (s)
V_s	Vessel service speed (knots)
∇	Submerged hull volume (m^3)
ϑ	Heel angle (degrees)

1. INTRODUCTION

The need to explore new approaches to assess ship stability in waves has led the IMO recent development of new stability criteria, named second generation intact stability criteria. The working group has decided to address the separate stability failure modes caused by the different following phenomena:

- Variation of righting arm in waves;
- Dead ship condition;
- Manoeuvring-related failures.

The ship compliance with relevant criteria is assessed by a multi-tired approach structured in three levels, with increasing accuracy of formulation: in case the ship is not able to comply with the 1st level criterion (L1), she has to be assessed according to 2nd level criteria (L2). A direct assessment (DA), for instance by means of a suitable numerical tool, should be carried out if, also at the 2nd level criterion, some vulnerability is evidenced. In case the issue can not be settled in the design phase, an Operational Guidance (OG) is to adopted, approved by the Flag Administration. Along the years, an intense research and development activity for each mode of failure has been carried out by the IMO Correspondence Group and by the international scientific community. An important and significant part of the literature to this regard is collected in the proceedings of the International Conference on Stability of Ship Ocean Vehicles (STAB) and the International Ship Stability Workshop (ISSW) of the latest years.

At the 2nd meeting of the SDC (it is the IMO Ship Design and Construction Sub-Committee, the new name of SLF Sub-Committee), in 2015, the final version of criteria for Parametric Roll (PR), Pure Loss of Stability (PLS) and Surf-Riding/Broaching (SR) has been finalized. While the complete criteria of Dead Ship condition (DS) and Excessive Acceleration (EA) failures are delivered at the end of the 3rd SDC session.

In this paper, specific attention will be paid to 1st levels of vulnerability of each stability failure, while the analysis of 2nd levels is limited to PLS and PR phenomena. Calculations carried out for the investigations in the field of Mega-Yacht vessels are based mainly on the rule text and guidelines available after the 4th SDC session (2017).

2. MODES OF FAILURE

In the following paragraph a short description of the stability failures taken into account at IMO is given. For a more accurate description on the physics of the phenomena see [1] and [2].

2.1 VARIATION OF RIGHTING ARM IN WAVES

The variation of ship righting arm is originated by the interaction between the ship and the wave and in this perspective the higher evidence is reached when waterline length is comparable with wavelength. Considerably important is also the wave steepness influence.

The ship immersed volume distribution and the shape of waterplane area are modified by the wave passage along the hull. Mega-Yachts, as well as, containerships or Ro-Ro ferries, are characterized by hull geometries with significant longitudinal and vertical variations, therefore they might present a high vulnerability to this phenomenon.

The variation of restoring moment in waves is the basis of two important stability failure modes:

- Parametric Roll;
- Pure Loss of Stability.

2.1 (a) Parametric Roll phenomenon

This phenomenon may occur when a ship is sailing in longitudinal regular seas and meets a train of waves with an encounter period of about half of the roll period.
With ship transversally far away from the upright position and the wave trough amidships, an increment of restoring moment is influencing the roll behaviour. When the ship comes back to the upright position, with a higher roll rate, because of the specific encounter frequency the wave crest has moved to amidships. In this condition, stability is decreased and the ship will roll to the opposite side with less resistance to heeling and greater speed. Then, once the ship reaches her maximum roll amplitude and the wave trough reaches again the amidships section, the stability increases again and the cycle starts with heel angle larger and larger.

2.1 (b) Pure Loss of Stability phenomenon

The wave steepness and the relationship between wave celerity and ship speed are very important parameters which may lead to PLS. As described above, the wave profile interaction with hull geometry implies a variation of waterplane area and a modification in submerged volume distribution. Taking into account a following sea, the worst condition occurs when a ship is reached by a wave having length and celerity comparable to the ship ones.

Due to the very low relative speed between ship and wave, a possible critical situation such as the wave crest amidships, can last enough time to jeopardize significantly the ship stability. Once the wave crest leaves amidships, the stability increases again increasing ship safety performances.

2.2 INSTABILITY IN DEAD SHIP CONDITION

In case of power loss in a storm, ships with a huge exposed windage area, such as Passenger ship and Mega Yacht, present the attitude to turn on beam seas. This behaviour is due to the combined action of wave drifting moment and aerodynamic wind moment. At first level, this phenomenon is modelled and assessed by means the so-called weather criterion.

The ship is considered while rolling under the action of a wind inclining moment. When the ship reaches the maximum windward angle it is supposed to suffer the effect of a wind gust. The action of the gust is assumed to last long enough to develop the heeling moment and to still insist on the ship while she has rolled to the other side completely. Whenever the angle of vanishing stability or the downflooding angle or 50° (whichever is the less) is achieved before the dynamic equilibrium angle, the stability of the ship is deemed as critical.

2.3 MANOEUVRING-RELATED INSTABILITIES

Manoeuvring-related instability includes the so-called Broaching-to. The phenomenon is a sudden and uncontrollable turn of the ship, sailing in following seas, despite a steering action to counteract the undesired event. This turning is usually accompanied by a large heel angle that may lead to capsizing. Surf-Riding is a phenomenon that usually precedes Broaching-to. This failure occurs when a wave from stern reaches the ship and accelerates it to the wave speed. During this situation, most ships are directionally unstable; so the uncontrollable turning may occur. Since Broaching-to and Surf-Riding phenomena are linked each other, it is possible to set the vulnerability criteria for Broaching by means of the probability that Surf-Riding occurs.

In order to assess a SR situation, the length and speed of the ship and wave are taken into account. SR rarely occurs for very long ships because waves with a length comparable to that of vessel are a lot faster than ship speed.

2.4 EXCESSIVE ACCELERATION FAILURE

This phenomenon occurs when a ship is rolling with a small roll period. Shorter roll period may lead to faster changes in linear velocity, therefore larger horizontal inertial forces are induced. Forces acting in a horizontal plane are more dangerous than vertical forces. The first one may produce losses of equilibrium, falls or damage to equipment or arrangement. While the vertical inertial forces cause brief overloading. The GM value has a significant influence on the roll period, in the way that larger metacentric height implies shorter roll period, with the consequence of higher accelerations.

3. THE RULE

In this section, the structure of criteria for each stability failure is described. A focus on the background that leads to the current formulation is reported in [3]. The complete text for PR, PLS and SR is delivered in [4] while DS and EA have been finalized in [5].

3.1 PARAMETRIC ROLL CRITERIA

The first vulnerability level analyses the hull performances observing the variation of *GM* in waves. The hull is supposed to be balanced in trim and sinkage on a set of defined waves. Each wave has the length equal to the ship length and a fixed wave slope value, equal to $S_w = 0.0167$. Positons of the crest along the hull are varied, the wave crest is to be centred amidships and subsequently at step $0.1 \cdot L$ forward and backward.
A vessel is not vulnerable to PR if:

$$\Delta GM_1 / GM_c \leq R_{PR} \quad (3)$$

where
R_{PR} is the standard as function of C_m and A_{bk}.
The second vulnerability criterion is made of two different analyses, *C1* and *C2*. The first one defines a set of 16 waves considering the wave crest amidships and at step $0.1 \cdot L$ forward and backward. It analyses both the influence of GM variation due to wave-hull interaction and the reference ship speed (V_{PRi}). The second criterion considers a wider range of waves, defined by a scatter diagram. A dynamic model is required to compute the maximum heel angle reached due to roll. Then, a ship is not considered vulnerable if:

$$max\{C1, C2\} \leq R_{PR0} \quad (4)$$

where $R_{PR0} = 0.06$.

3.2 PURE LOSS OF STABILITY CRITERIA

For this phenomenon, the following criteria should not apply to vessels with a Froude number at service speed lower than 0.24. The first level analyses the ship in terms of the minimum *GM* in waves. The hull is considered to be balanced in sinkage and trim on a set of waves, with the same length and a constant wave slope $S_w = 0.0334$. The wave crest is to be centred amidships and at step of $0.1 \cdot L$ forward and backward. A ship complies with the first vulnerability level when:

$$GM_{min} \geq R_{PLA} \quad (5)$$

where $R_{PLA} = 0.05$ (m).

The second vulnerability criterion is to be applied if the the ship doesn't comply with the first one. Also in this case, the second level analyses two different parameters to assess the ship that is considered to be statically positioned in waves, i.e. *CR1* and *CR2*. The first analysis assesses statically the vanishing angle of stability, while the second one evaluates the static heel angle reached by the ship under the effect of a heeling lever, as function of wave characteristics. For both criteria, the vessel is considered to be balanced in sinkage and trim on a huge set of wave extrapolated from a scatter diagram. For each wave, the crest is to be centred amidships and at step of $0.1 \cdot L$ forward and backward. Then, a ship is considered not to be vulnerable if:

$$max\{CR1, CR2\} \leq R_{PL0} \quad (6)$$

where $R_{PL0} = 0.06$.

3.3 SURF-RIDING / BROACHING CRITERIA

A vessel is considered vulnerable to the first vulnerability level if its length is greater than 200 (m) or the Froude number at service is equal or lower than 0.3. A more specific and detailed analysis of this phenomenon is carried out within the second vulnerability level, which is not addressed by this investigation.

3.4 DEAD SHIP CONDITION CRITERIA

The first vulnerability level for DS consists of the application of the so-called Weather Criterion issued by IMO within [6]. A vessel is not vulnerable to DS if:

$$a \leq b \quad (7)$$

where a is the area on the left side of the equilibrium angle between the wind lever and GZ; b is the area on the right side of the equilibrium angle between GZ and the wind lever. The wind lever is defined by the rule as function of exposed lateral area. In Figure 1, a graphical description of the requirements of the standard is shown.

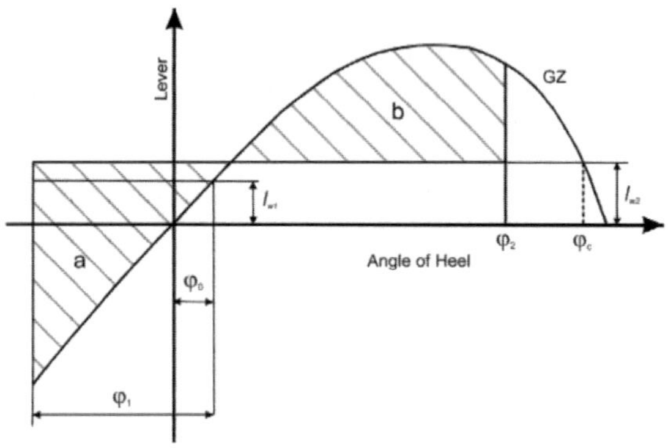

Figure 1: Weather criterion.

3.5 EXCESSIVE ACCELERATION CRITERIA

The assessment of this failure is based on a statistical approach to describe the external force that develops large roll motion. Within the first level, this method is simplified by means some approximation described in the Annex 7 of [2]. The final formulation defines that a ship is not vulnerable if:

$$\phi \cdot k_L \cdot (g + 4\pi^2 \cdot h / T^2) < R_{EA1} \qquad (8)$$

where R_{EA1} is the standard to be defined between 5.3 and 8.9 (m/s^2); ϕ is the characteristics roll amplitude (rad); k_L is a factor taking into account action of roll, yaw and pitch; h is the height (m) where passenger or crew may be present on board the ship and T is the rolling period (s).

4. APPLICATION CASE

SGISC are mainly focused on merchant vessel as commonly meant, i.e. cargo and passenger ships; although some Mega-Yacht units have been considered in the different applicability analysis of the criteria (two vessels in [7], nine units in [8] and thirteen in [9]) they were few compared to the rest of the ship typologies investigated in the recent years by the community.
The main dimensions and characteristics of the hull considered in this paper in zero trim conditions are indicated in Table 1.

Table 1: Main Data

Original Model – M-0		
L_{OA}	74,39	(m)
L	66,69	(m)
B	13,20	(m)
d	3,30	(m)
∇	1.589	(m^3)
L/B	5,28	(-)
B/d	4,00	(-)
V_s	17,2	(kn)
GMT	1,716	(m)
C_B	0,547	(-)
C_X	0,918	(-)
T_{roll}	8,793	(s)

The ship natural roll period is calculated according to [6]. The vertical position of center of gravity is KG = 5,70 (m) while the vertical position of center of buoyancy is KB = 1,97 (m).

To enlarge the investigation domain, four additional variants are derived, starting from the main hull. The L/B ratio has been increased and decreased by 10% and 20%, keeping constant the length. Also KG is kept fixed. Main data of new models are listed in Table 2.

Table 2: Data of different models

	M-20	M-10	M+10	M+20	
L/B	4,22	4,75	5,80	6,33	(-)
B/d	5,00	4,44	3,64	3,33	(-)
B	16,50	14,67	12,00	11,00	(m)
∇	1.986	1.765	1.444	1.324	(m^3)

The EA criterion requires the definition of height of the location were crew or passenger may be present. It is fixed at 12 (m) from keel line which is the position of the wheelhouse. For a larger investigation another value has been also selected i.e. 9 (m).

The objective of this paper is to achieve the knowledge of how a design modification may influence the vulnerability to stability failures. A modification in breadth has been done to understand how results might change assessing different ship geometries but with the same wave data set. As described in regulation, the wave characteristics depend on the length of the ship to assess. The influence of length on the vulnerability to stability failure was already addressed in [10].

4.1 COMPUTATIONAL TOOLS

The ship righting arm in wave computations are carried out by means of an "in house" computational tools, developed at the University of Genoa (UNIGE). It considers the hull statically positioned on a wave crest taking into account the equilibrium in sinkage and trim. The calculation process may be divided in three parts: the first defines input data, the second calculates geometric characteristics while the third analyses stability parameter. For a specific explanation how the computational code works see [11] and [12].

	Parametric Roll	Pure Loss of Stability	Surf Riding / Broaching	Dead Ship Condition	Excessive Acceleration
1st Level	available	available	available	available	available
2nd Level	available	available	under development	under development	
Direct Assessment					

Figure 2: Codes at University of Genoa.

For each phenomenon, specific computational codes for vulnerability assessment for the two vulnerability levels have been developed. They implement the rule text and they evaluate the vulnerability as function of different KG and draft. At present, the computational codes modules already available for UNIGE are shown in Figure 2.

5. RESULTS AND COMMENTS

Hereinafter results are shown as calculated for each model and each stability failure, applying the relevant criteria at the first vulnerability level. In Figure 3, for each failure, the KG_{max} limit curves of the original hull (M-0) are shown. On the vertical axis the values of KG are reported, whit drafts on horizontal axis. It is to be noted that, in the case of EA, unlike other phenomena, the vessel is considered not vulnerable when KG is above the relevant curve.

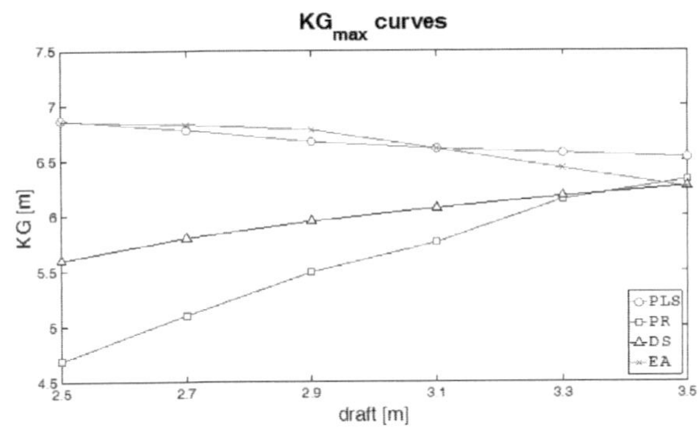

Figure 3: KG_{max} limit curves as function of draught for the original M-0 vessel.

With reference to Figure 3, within the set of the observed drafts, it appears that it is impossible to design a not vulnerable ship. In fact, EA curve is above the other curves and this imply an empty domain of solutions. As mentioned above, the EA phenomenon depends also on the height of the location assessed. Beside a height of 12 (m), also 9 (m) above keel line is assessed to evaluate the influence of this factor. Moreover, since the standard for EA is at present an option between 5.3 (m/s^2) and 8.9 (m/s^2), in Figure 4 the KG_{max} limit curves computed with different height and standards for EA are given.

It is evident that a change in the assessment point has an influence on the results and on the ship feasibility, but this is effective only in relation with the standard value. In fact, the change of the height from 12 (m) to 9 (m) doesn't produce any practical improvement when analysing results in relation with the standard of 5.3 (m/s^2). On the contrary, when considering 8.9 (m/s^2) as standard, a huge difference can be observed that enables the design of a feasible ship for all the range of analysed drafts. The area of acceptable KG values is, in fact, identified between the PR and EA relevant curves. This interesting behaviour is further investigated and in Figure 5 it is possible to observe directly, the influence of the standard values coupled with the investigated heights. A ship is not vulnerable if the criterion is lower than the standard considered.

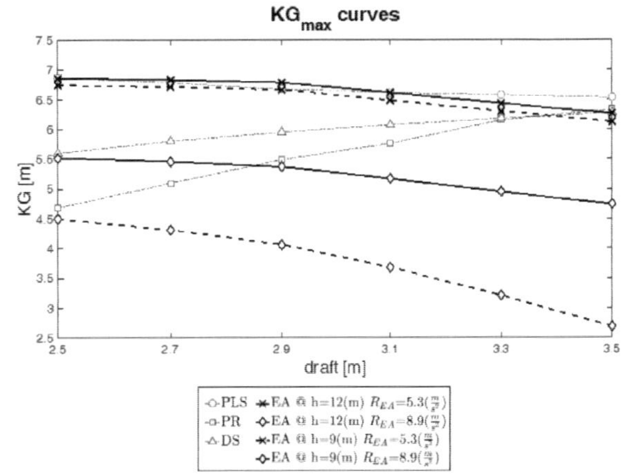

Figure 4: KG_{max} limit curves with different standard and height of location for EA.

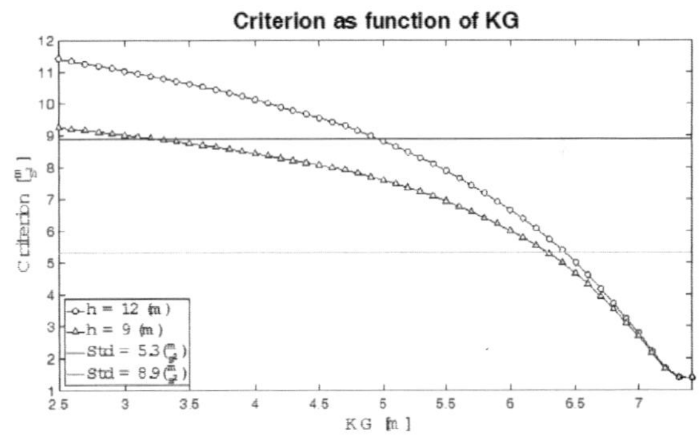

Figure 5: Influence of investigated height at $d = 3.3$ (m)

Due to the non linear behaviour of the outcome curves, the intersection points with the 5.3 (m/s^2) standard are much closer than the intersection point with the 8.9 (m/s^2) standard.

To assess the influence of a change in ship breadth, a comparison between different models (as described above) for each mode of failure is carried out. The following results are referred to a constant draft equal to d = 3,3 (m). In Figure 6, the KG_{max} limit curves are shown as function of breadth.

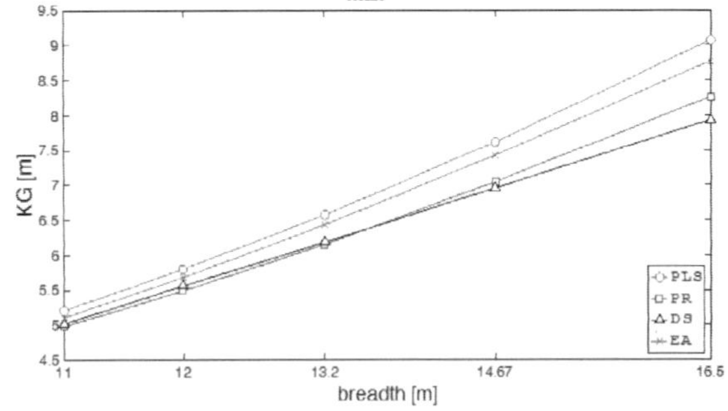

Figure 6: KG_{max} limit curves as function of breadth.

All the curves present a very similar behaviour i.e. an improvement of performance while breadth is increasing. This is in line with what expected, but it is possible to appreciate such influence on a quantitative basis.

Further analyses pointed out that a lack of consistency is possible between first and second level criteria of PLS, as shown in Figure 7. In fact, for all models, the curve for second level is lower than the first one and this is not in the philosophy of the multi-layered approach. However, for the specific case of the PLS, the trend of both curves is similar and it slightly decreases with an increment of draft. This shortcoming, can be overcome, for example acting on the value of the standard, i.e. the threshold for acceptability.

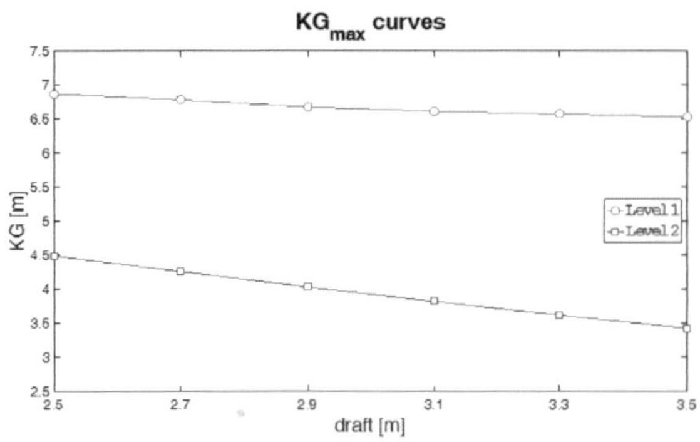

Figure 7: Level 1 and Level 2 of PLS

The situation is more critical as far as the PR is concerned. In fact, we find again an expression of inconsistency between the two levels, even though only for higher drafts. Unfortunately, in this case it is unlikely that the shortcoming can be fixed only acting on the standard value. The two curves have in fact opposite behaviour, i.e. one is increasing while the other is decreasing. The comparison between Level 1 and Level 2 results is reported also for PR in Figure 8.

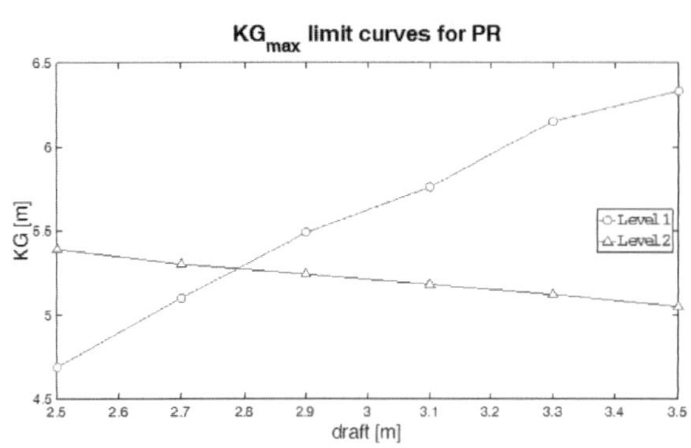

Figure 8: Level 1 and Level 2 of PR

6. CONCLUSIONS

Within the framework of Second Generation Intact Stability Criteria, the application of the first levels of vulnerability criteria for all the different stability failure modes, is investigated. As far as, PR and PLS, also the application of second levels is carried out.

The aim of the activity is to understand how a design modification in an early design phase might influence the susceptibility to the dynamic instability of vessel in seaway. In order to do this, a Mega-Yacht hull has been modified varying the *L/B* ratio, acting only on breadth, with constant length.

Results show an important effect of breadth change and a quantification of ship stability improvement is evidenced thanks to maximum *KG* curve plot. The effect of a change in geometry in terms of breadth was expected but in the investigation it has been possible to point out the influence on a quantitative base.

It is recognized as well that some of the failure modes and the relevant criteria imply the influence of some other physical parameter that might be object of future investigation.

Observing the results of the present study, some possible inconsistencies have been evidenced as well, in relation with the first and the second vulnerability level, as regards PLS and PR. In fact, in principles, the first level KG_{max} curve should always be lower than second level one and this never happens for the PLS and for PR it is respected only for lower drafts.

An other important comment is about the EA criterion: it might happen that the KG curve obtained with the first level criterion, combined with the other stability failure mode curves, eliminates any possible design domain for the ship, for the whole draft range that is taken in to account. A further analysis on this aspect has evidenced that the selection of the proper standard can fix this problem.

7. REFERENCES

1. BELENKY, V., DE KAT, J.O. & UMEDA, N., 'Toward Performance-Based Criteria for Intact Stability', *Marine Technology 45*, 2008.

2. BELENKY, V., BASSLER, C.C. & SPYROU K.J., 'Development of Second Generation Intact Stability Criteria', *Hydromechanics Department Report (Naval Warfare Center Carderock Division - USA)*, 2011.

3. UMEDA, N., 'Current Status of Second Generation Intact Stability Criteria Development and Some Recent Efforts', *Proceedings of the 13th International Ship Stability Workshop*, 2013.

4. IMO, 'Development of Second Generation Intact Stability Criteria', *SDC 2 / WP.4*, 2015.

5. IMO, 'Finalization of Second Generation Intact Stability Criteria', *SDC 3 / WP.5*, 2016.

6. IMO, 'Adoption of the International Code on Intact Stability', *MSC Resolution MSC.267(85)*, 2008.

7. IMO, 'Sample calculation results of the proposed second-generation intact stability criteria regarding the Levels 1 and 2 failure modes Parametric Roll and Pure Loss of Stability for Superyachts', *SDC 1-5-8*, 2013.

8. IMO, 'Information collected by the Correspondence Group on Intact Stability regarding the second generation intact stability criteria development', *SDC 1 - INF.8 – Annex 9*, 2013.

9. IMO, 'Information collected by the Correspondence Group on Intact Stability regarding the second generation intact stability criteria development', *SDC 3 – INF. 10 – Annex 7*, 2015.

10. PETACCO, N. & GUALENI, P., 'Second Generation Intact Stability criteria for mega-yachts: application and design considerations', *3rd International Conference on Maritime Technology and Engineering*, 2016.

11. CORADDU A., GUALENI, P. & VILLA, D., 'Investigations about wave profile effects on ship stability', *XVI Congress of International Maritime Association of the Mediterranean*, 2011.

12. CORADDU A., GUALENI, P. & VILLA, D., 'Vulnerability assessment for the loss of stability in waves: some application cases for a further insight into the problem', *Proceeding of 11th International Conference on the Stability of Ships and Ocean Vehicles*, 2012.

8. AUTHORS BIOGRAPHY

Paola Gualeni is graduated in Naval Architecture and Marine Engineering. She is Associate Professor at the University of Genoa and she teaches Ship Hydrostatics and Stability. Her main fields of interest are Ship Safety and Innovative Ship Design Methodologies.

Petacco, N. is graduated in M.S. Naval Architecture and Yacht Design at the University of Genoa in 2014. He is a Ph.D. student in Naval Architecture at the University of Genoa since November 2015. His main areas of interest are ship safety and stability concerning the dynamic effect of hull and wave interaction, in particular for mega yacht units.

FLOODABLE DOCKING GARAGES

C. Barba Fau, Lloyd's Register EMEA, Italy

SUMMARY

One of the latest trends in the large yacht world is, for the garage compartment, to feature a floodable dock. This is preferable to Owners as this avoids the installation of travelling cranes on the ceiling of the lower deck for the embarkation/disembarkation of tenders and other small recreational watercraft and also reduces significantly the total weight of the vessel.

With this paper Lloyd's Register aims to analyse the feasibility of "floodable docking garages" fitted on yachts, either in open or closed door scenarios, and to provide guidance on the design of this type of arrangement, when dealing with structural, load line and stability aspects, in order to ensure that an acceptable level of safety is achieved in compliance with the principles of the Lloyd's Register Rules and Regulations for the Classification of Special Service Craft (hereinafter "SSC Rules") and statutory requirements being the most common the MCA Large Commercial Yacht Code (hereinafter "LY3 Code").

1. INTRODUCTION

In the last years several designers/shipyards have required the feasibility of fitting floodable docks in the garage compartment of the vessel.

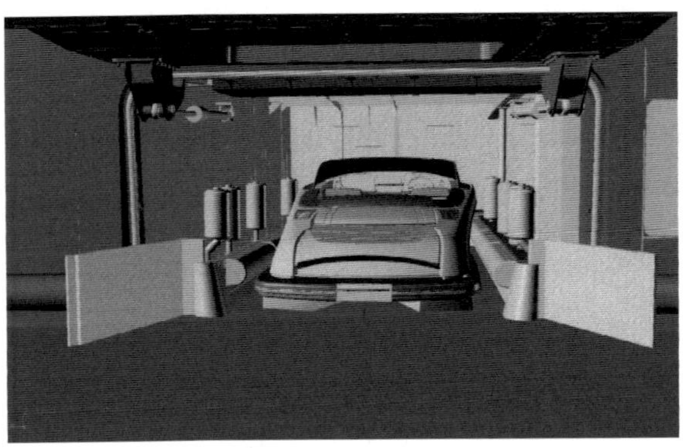

Figure 1: Floodable docking garage, MY "J'ade"

This type of arrangement involves that the sill height of the shell door of such compartment will be fitted below the design waterline.

While in larger yachts it might be possible to consider this compartment as non-buoyant, on yachts with rule length lesser than 50 meters this is not always possible because the yacht may not comply with the intact and damage stability requirements.

To fulfil an adequate level of safety, a deeper investigation of the arrangement in subject has been carried out, and with this paper LR aims to achieve a solution which could satisfy all the parties involved in the design, construction, surveillance and the recreational and operational needs of the Owner (easy operability of the crafts stored inside, less training for crew operating cranes, less crew involved in the operations, fast availability of the crafts).

2. TYPICAL REGULATORY FRAMEWORK

As stated on regulation 21(2) of the International Load Line Convention (ILLC), the minimum sill height of shell doors must be 230 mm from the upper edge of the uppermost load line. Otherwise additional fittings need to be provided to maintain the watertight integrity of the vessel.

Based on the above Lloyd's Register (LR) requires, on the SSC Rules, a minimum sill height of 600 mm above the design waterline or at 300 mm might be allowed if the doors of the compartment giving internal access are provided with a minimum sill of 600 mm height from the design waterline.

The LY3 Code, requires also to have a minimum sill height of 600 mm and when this is not possible due to the design of the craft a reduction might be considered by the Administration.

3. ASSUMPTIONS

Fitting of a floodable docking garage, means having a shell door with a sill height below the design waterline and ingress of water into the vessel when the door is open during the manoeuvring of the tender or recreational small crafts (see Fig.2).

Hence, requirements of Part 3 Ch.4 Sec. 4 of the SSC Rules and Section 4.4.3.1.1 of the MCA LY3 Code, relevant to the minimum sill height of the shell door from the design waterline, cannot be complied with. Moreover, the watertight integrity of the vessel might be compromised.

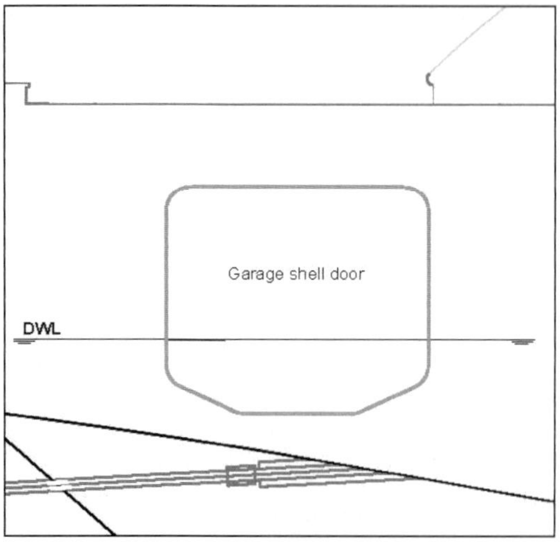

Figure 2: Side view of motor yacht A

4. POSSIBLE FORESEEABLE SCENARIOS

To maintain the watertight integrity of the vessel by limiting the amount of water, which can accidentally floods the garage space, and preventing the effect of this flooding on stability the following scenarios related to the shell door giving access to the garage compartment need to be evaluated:

4.1 SHELL DOOR IN CLOSED POSITION

When the vessel is at sea, generally shell doors will be closed, therefore the ingress of water into the garage space might be due to:

- The leakage from the side shell door gasket arrangement.
- The heeling, trim and increase of draught of the vessel (due to damage in another watertight compartment) plus a possible leakage from the shell door gasket arrangement.

4.2 SHELL DOOR IN OPENED POSITION

When the vessel is in harbour or sheltered waters conditions shell doors might be opened, consequently the following may occur:

- Ingress of water into the dock recess.
- Additional ingress of water into the garage space due to an accidental wave, e.g waves caused by the transit of another vessel during the manoeuvring of the tender and/or other recreational small crafts.

5. CRITICAL ASPECTS

LR, when assessing subject arrangements, is mainly focused in evaluating the aspects related to the stability, Load Line and strength for the space.

5.1 STABILITY

It is the view of Lloyd's Register that "floodable docking garage" space can be considered buoyant if all the following are fulfilled:

- The shell door is watertight, and "floodable docking garage" space is bounded by watertight bulkheads without any openings to separate the compartment from any other compartments below the freeboard deck.
- The shell door will not be opened when the yacht is under seagoing conditions, it can only be opened when the yacht is in harbour conditions or at anchor in sheltered waters.
- A drain tank or other means of dumping water from the garage space will be provided.
- As required on Section 4.4.3 of the MCA LY3 Code, the effect of flooding the garage dock on stability will be considered.
- Guidance for the Master should be included in the Stability Booklet to highlight the importance of the watertight integrity of the garage door on the stability aspects of the yacht.

Otherwise, the garage compartment needs to be considered non-buoyant.

5.2 LOAD LINE

The "floodable docking garage" compartment needs to be considered as a recess on the freeboardeck.

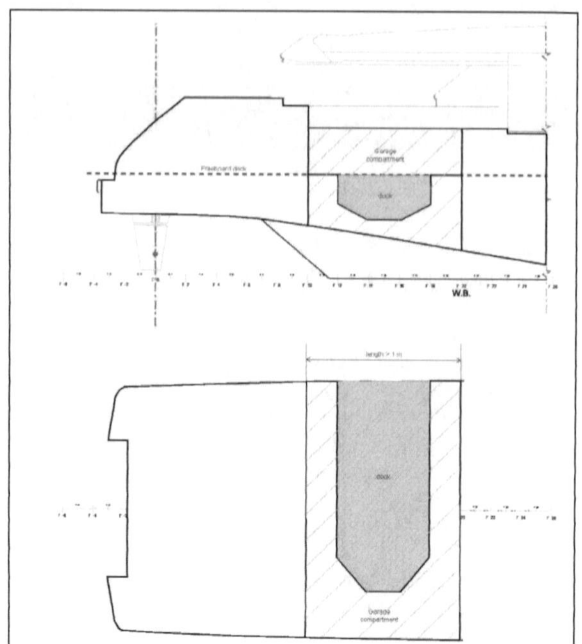

Figure 3: Side and top view of floodable docking garage of motor yacht A.

Since the garage compartment extends to the sides of the ship and is in excess of 1 metre in length, the freeboard deck of the vessel will be considered discontinuous. As indicated on Regulation 3(9)(c) of the ILLC), the lowest line of the exposed deck (garage deck) and the continuation of that line, parallel to the upper part of the deck, is to be taken as freeboardeck.

Moreover, the dock volume shall be considered as volume for loss of buoyancy.

5.3 STRENGTH

The garage boundaries need to be design watertight and the dock boundaries will be assumed equivalent the hull shell.

6. EQUIVALENT ARRANGEMENT

6.1 MINIMUM RECOMMENDATIONS

The "floodable docking garage" deck should be of a watertight design and generally fitted above the design waterline.

The floodable dock recess should not extend side to side and should be fitted within a watertight compartment.

In any case the shell door will be closed before leaving the port. The door might be opened only in calm water in protected areas, in no case the door will be opened when the sea conditions are not suitable for the embarkation/disembarkation of the tender and/or small recreational crafts, or in open seas. It is the Master responsibility to decide whether the conditions are suitable or not.

6.2 ADDITIONAL FITTINGS

Doors of the "floodable docking garage", leading to the internal compartments of the yacht, should be sliding, watertight and with a minimum sill height of 600 mm above the maximum design waterline as required on Part 4 Ch.2 Sec.4 of the SSC Rules and Sec.4.3.1 of the MCA "LY3" Code. When the sill height cannot be achieved due to the arrangement and design of the vessel, adjacent space will be bounded watertight and will be considered together with the garage and doors from this space giving internal access should be provided with a sill of 600 mm height above the maximum design waterline in order to guarantee an equivalent arrangement

A second barrier of equivalent strength and watertightness of the shell door need to be fitted

A spare gasket for the shell door and the second barrier will be stored on board in case of leakage

The garage deck should be watertight to provide a reserve of buoyancy when damage is caused on another watertight compartment of the vessel and avoid ingress of water in sensitive areas of the compartment (e.g. rudder actuators).

The floodable dock recess should be fitted with a minimum of 600 mm coaming around its perimeter on the garage deck. Moreover, as required on the MCA LY3 Code "*any recess is to be able to be drained in 3 minutes and at an angle of 10 degrees for motor yachts*".

A water leakage detection system with audible alarms shall be arranged in way of side shell door and the secondary barrier to provide indication to the navigation bridge and to engine control room for any leakage through the door and the barrier. Furthermore, a drainage system should be provided to guarantee the freeing of the water in the space within the shell door and the secondary barrier.

Flood detection system and suitable drainage arrangement shall be provided on the "floodable docking garage" compartment

Flood, leakage, and drainage detection systems shall be connected to the Main and Emergency switchboard.

6.3 ADDITIONAL ACTIONS

Suitable instructions should be posted by the Designer/Shipyard to the Master relating to the immediate actions to be taken in case of ingress of water into the garage space, when the door is open.

Watertight doors giving access to the adjacent spaces shall be closed at all times while the yacht is at sea with a suitable notice affixed to the door and an alarm with an indicator and control on the bridge.

7. CONCLUSIONS

Along this paper, the possible problems that this type of arrangement could cause to the safety of the vessel have been described. Furthermore, it has been also explain how to reach an equivalent arrangement aiming at the same level of safety that could satisfy SSC Rules and statutory requirements, being the most common the MCA "LY3" Code, having in mind the needs and requirements of the Owner.

However, since the "floodable docking garage" arrangement is subject to statutory requirements the final acceptance of the arrangement is to be given by the Flag Administration.

8. DISCLAIMER

Working together for a safer world

Lloyd's Register and variants of it are trading names of Lloyd's Register Group Limited, its subsidiaries and

affiliates. Lloyd's Register EMEA (Reg. no. 29592 R) is a registered society under the Co-operative and Community Benefit Societies Act 2014 in England and Wales. Registered office: 71 Fenchurch Street, London, EC3M 4BS, UK.
A member of the Lloyd's Register group.

Lloyd's Register Group Limited, its affiliates and subsidiaries and their respective officers, employees or agents are, individually and collectively, referred to in this clause as 'Lloyd's Register'. Lloyd's Register assumes no responsibility and shall not be liable to any person for any loss, damage or expense caused by reliance on the information or advice in this document or howsoever provided, unless that person has signed a contract with the relevant Lloyd's Register entity for the provision of this information or advice and in that case any responsibility or liability is exclusively on the terms and conditions set out in that contract.

9. ACKNOWLEDGEMENTS

The author would like to thank her colleagues for sharing their knowledge and experience and to Lloyd's Register for the support and the opportunity given.

10. REFERENCES

1. International Maritime Organisation, 'International Convention on Load Lines, 1966 and its 1988 Protocol (including 2003 and 2004 Amendments)', IMO, 2005.

2. Lloyd's Register Rules and Regulations for the Classification of Special Service Craft, July 2015.

3. Marine and Coastguard Agency, 'The Large Commercial Yacht Code (LY3)', Marine and Coastguard Agency, August 2012.

11. AUTHORS BIOGRAPHY

Cristina Barba Fau currently holds the position of Specialist to Lloyd's Register, at Trieste Technical Support Office, dealing with structural and load line aspects of Yachts and Mega Yachts. She joined LR in 2008. Her previous experience includes design, construction, repair and refitting of sailing vessels and motor yachts. Mrs Barba Fau is an Associate Member of the Royal Institution of Naval Architects.

NUMERICAL TOOLS AND EXPERIMENTAL ACTIVITIES FOR THE PREDICTION OF NOISE PROPAGATION ON BOARD SUPERYACHTS

A. Tonelli, D. Torre – RINA Services - Italy
G. Vergassola, D. Boote – University of Genova - Italy

SUMMARY

The Italian Classification Society, RINA S.p.A., is developing a new set of Rules regarding comfort conditions on board large motor-yachts. According to present owner requirements and technical enhancement, the new limit values will be more restrictive especially for yachts with more than 65 meter in length. In this perspective designers and shipyards should consider structure borne noise and vibration level as leading parameters since the earliest design stages.

In this paper some numerical tools, which allow to predict noise propagation on board before sea trials are carried out, are presented. The reliability of the proposed tools has been tested by means of experimental activities for the evaluation of damping loss factor on the viscoelastic materials employed in different parts of the vessel such as floating floors and engine foundations.

NOMENCLATURE

ABN	Air-borne noise
DLF	Damping loss factor
f_c	Coincidence frequency [Hz]
FEA	Finite Element Analysis
FEM	Finite Element Method
NVH	Noise, Vibration, Harshness
L_{OA}	Length Over All [m]
SBN	Structure-borne noise
SEA	Statistical Energy Analysis
SPL	Sound Pressure Level [dB]
η_i	Damping loss factor in SEA matrix
η_{ij}	Coupling loss factor in SEA matrix
ω	Radiance frequency (rad)

1. INTRODUCTION

In the last years, the main parameter that designers have to face during a megayacht project has been shifted from velocity to human comfort on board, in order to fulfil owners' requirements.

In this perspective, the assessment of level of noise and vibration has become a necessary task that have to be carried out even during the early design stages, when modifications in the structural layout and/or in the insulation plans can be accomplished with low working times and costs.

Moreover, Classification Societies are developing more stricter rules in terms of NVH, by fixing new limit values for vibration velocities and sound pressure level (SPL) especially for super and megayacht (L_{OA} > 65 m) [2].

For what vibration assessment is concerned, numerical procedures based on FEA are well-established in literature [3] [4]. The numerical evaluation of air-borne noise (ABN) and structure-borne noise (SBN) is emerging only in the last few years by the development of new software based on the Statistical Energy Analysis (SEA). By mean of this procedure, the SPL distribution can be calculated even for those high frequency ranges, where deterministic solution cannot be considered reliable.

In this paper, after an overview of the SEA for marine application, numerical tools and related experimental activities are presented, in order to evaluate and therefor reduce the uncertainties and, so, to increase the reliability of SEA. In particular, the two main troubling parameters (i.e. input power of sources and the damping loss factor of structural components) have been deepened in this work.

2. RINA ADDITIONAL COMFORT NOTATIONS

Mandatory regulations as "The Code on noise levels on board ships" (hereinafter referred to as "the Code") has been developed by IMO [1] to provide international standards for protection against noise and to promote "hearing saving" environments on conventional passenger and cargo ships in order to avoid a noise-induced hearing loss.

The Code is intended to provide the basis for a design standard to prevent the occurrence of potentially hazardous noise levels and to provide an acceptable environment for seafarers with compliance based on the satisfactory conclusion of sea trials that result in issuance of a Noise Survey Report which show the noise to a level which is not potentially harmful

On the way of the IMO, classification Societies have developed the so called COMFORT class notations with more strictly requirements and recommendations not only to protect the people on board from excessive noise levels but to provide living spaces with an acceptable degree of comfort in navigation, and as quiet as possible, at rest also with verification of acoustic insulation between accommodation spaces and impact noise on floor above cabins.

The Additional voluntary notation COMFORT class is assigned only when measurements at sea trial highlights that noise and vibration limit levels are respected.

RINA's additional class notations for comfort on board can in turn be assigned depending on the type of vessels as follows:

- COMF-NOISE to assess the noise level on all types of ships (compliance with Code of Noise Levels on Board Ships, IMO Resolution MSC.337(91))
- COMF-VIB to assess the vibration level on all types of ships
- COMF Yacht to assess the noise and vibration levels on pleasure or charter yachts
- COMF (LY) to assess the noise and vibration levels on large yachts (Lpp>60m)
- DOLPHIN to set limit for the underwater radiated noise for both commercial vessels, pleasure yachts and yachts (above 24 meters).

The first three notations have been developed with prescrittive limits obtained by numerous test on existing vessel in collaboration with the shipyards and shipowner in order to provide reference values in compliance with the international standards.

Keeping into account results of research projects in which several underwater radiated noise measurements have been carried out, DOLPHIN could be adopted in areas highly sensible to environmental issues like areas of marine mammals reproduction.

The DOLPHIN notation is not addressed directly to the noise levels inside but the restriction on the noise emitted underwater strongly influence the internal noise and the analyses can be developed with SEA methodology adopted for the internal noise.

The notation COMF (LY) is the last one developed with the intention to obtain the best compromise between the limit levels of a passenger ship with reduced vessel length and the performance required to luxury yacht taking into account also the wellness of the crew.

In the following tables the limit levels for noise comfort on large yachts are shown:

Table 1: Noise insulation characteristics of the divisions

Division between	**R'w (dB)**	
	A level	**B level**
Passenger cabins/cabin (1)	43	35
cabin/corridor	40	35
Cabin/stairway	48	45
Cabin/public spaces (excluding corridors, stairway and discotheque)	55	50
Cabin/galley	40	35
Disco/cabin	65	55
Cabin/machinery room	55	50
Crew cabin/crew cabin	35	35
Crew cabin/corridor	33	27
(1) Except for Cabin suite / cabin with communicating door where the R'_w limits are reduced by 3 dB		

Table2 : impact noise insulation characteristics of floor

Floor above cabin	**$L'_{n,w}$ (dB)**
Soft floor	50
Hard floor	55
Dance floor, stages, gymnasium floor	45
Floor above public spaces	**$L'_{n,w}$ (dB)**
Soft floor	50
Hard floor	45
Dance floor, stages, gymnasium floor	55

Table 3: Noise limits levels – passenger spaces – navigation

Type of space	**Navigation**	
	L_A	**L_B**
Passenger spaces		
Passenger cabins – superior	50	55
Passenger cabins – standard	53	60
Public spaces - lounges	55	60
Open main deck recreation areas, aft	75	79
Open deck recreation areas above, aft **(1)**	65	70
Wheelhouse	60	75
(1) +10 dB(A) if less than 1 m from the ventilation inlet/outlet		

Table 4: Noise limits levels - passenger spaces - At berth

Type of space	At berth	
	L_A	L_B
Passenger spaces		
Passenger cabins – superior	45	50
Passenger cabins – standard	48	53
Public spaces - lounges	50	55
Wheelhouse	50	65

Table 5: Noise limits levels - crew spaces – navigation

Type of space	Navigation	
	L_A	L_B
Crew accommodation		
Crew cabins	55	60
Senior officer cabins	52	55
Navigation spaces/radar room	65	65
Radio room	60	60
Look-out, incl. Navigation wings and windows	70	70
Hospital	55	60
Public crew spaces	60	65
Work spaces without equipment operating	70	75
Workshop other those forming part of machinery	85	85
Offices	60	65
Mess room/recreation room	60	65
Engine control room	75	75
Crew open deck **(1)**	75	75
(1) +5 dB (A) if less than 1 m from the ventilation inlet/outlet		

Table 6: Noise limits levels - crew spaces - At berth

Type of space	At berth	
	L_A	L_B
Crew accommodation		
Crew cabins	50	55
Senior officer cabins	50	55
Mess room/recreation room	55	60
Engine control room	70	75

The notation is differentiate by a letter A or B which represents the merit level achieved for the assignment of the notation, the merit A corresponding to the lowest level of noise and vibration.

In the case of different levels obtained for passenger and crew, the COMF LY PACR notation with two letters will be assigned. The first will indicate the acceptance criteria for passenger accommodation, whilst the second will indicate the crew comfort criteria.

At design levels verification of compliance could be addressed on SEA calculation with the target to obtain conservative results different respect to the limits by a margin related to the uncertainties of the methodology.

3. THE STATISTICAL ENERGY ANALYSIS FOR MARINE APPLICATION

3.1 THEORETICAL BACKGROUND

The SEA is a framework of study [5] for the prediction of sound and vibration for complex dynamic systems. The statistical aspect of this method emphasizes the difference with deterministic numerical approaches (such as FEM, FDM, FVM, etc.) that are not able to predict the correct behaviour for high frequency ranges in which the mode shape and resonance frequencies highlights great sensitivity to small details of geometry and construction.

This aspect could be solved in traditional FEA by using the so called Monte Carlo Approach [6] with a great increase in computational time and costs.

The Statistical Energy Analysis from a pure theoretical point of view is a set of power balance equations, derived from the principle of conservation of energy.
Considering that a complex dynamic system could be divided into at least two simpler subsystem (Fig. 1) the set of power balance equations can be written as:

$$\begin{cases} \sum_1^2 \pi_{in,i} = \sum_1^2 \pi_{out,j} \\ \pi_{out,i} = \pi_{ij} + \pi_{diss,i} \end{cases} \qquad (1)$$

Where $\pi_{in,i}$ is the power input in the i-th subsystem, $\pi_{out,i}$ is the power output, π_{ij} is the amount of power that is transmitted between i-th and j-th subsystem and $\pi_{dissipated,i}$ is the power that is dissipated by the i-th subsystem.

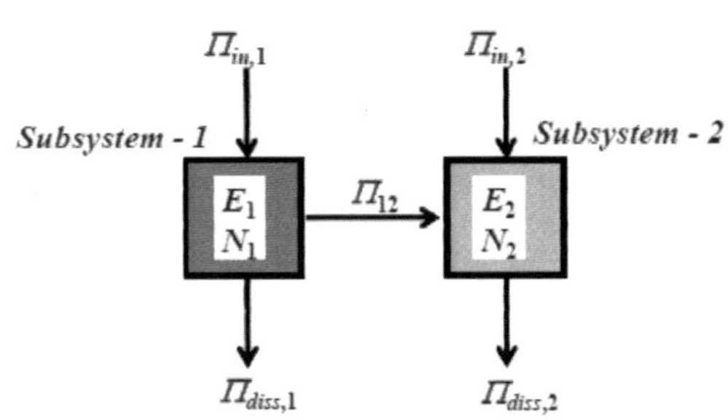

Figure 1: SEA system composed by 2 subsystems

In Fig. 1, the term E_i denotes the total energy of each subsystem and N_i is the modal density.

The dissipated power of each subsystem can be written as:

$$\pi_{diss,i}(\omega) = \omega \cdot \eta_i \cdot E_i \quad (2)$$

where η_i describes the damping loss factor (DLF) of the subsystem.

The last unknown term is the transmitted power that is depending on the energy of each subsystem as it follows:

$$\pi_{ij}(\omega) = \omega \cdot \eta_{ij} \cdot n_i \cdot \left(\frac{E_i}{n_i} - \frac{E_j}{n_j}\right) \quad (3)$$

By applying these sets of equations, a subsystem that could be decomposed in millions of nodal degrees of freedom in a deterministic numerical analysis, in SEA has only 1 energy degrees of freedom, so the calculation is really less heavy.

The coupling loss factor η_{ij} depends on the type of subsystem (e.g. beam, shell, acoustic cavity) and on the transmission path (mass law, resonant, non resonant or double wall) under consideration.

Under these consideration, a SEA problem for a dynamic system composed by two subsystem could be written in matrix form as:

$$\begin{bmatrix}\pi_{in,1}\\ \pi_{in,2}\end{bmatrix} = \omega \cdot \begin{bmatrix} n_1(\eta_1+\eta_{12}) & -n_1\eta_{12} \\ -n_2\eta_{21} & n_2(\eta_2+\eta_{21})\end{bmatrix} \begin{bmatrix}\frac{E_1}{n_1}\\ \frac{E_2}{n_2}\end{bmatrix} \quad (4)$$

Generalizing Eq. 4 for a N-components dynamic system, the SEA matrix system is:

$$\begin{bmatrix}\pi_{in,1}\\ \vdots \\ \pi_{in,N}\end{bmatrix} = \omega \cdot \begin{bmatrix} n_1(\eta_1+\sum_{i\neq 1}\eta_{1i}) & -n_1\eta_{12} & \cdots & -n_1\eta_{1N} \\ -n_2\eta_{21} & n_2(\eta_2+\sum_{i\neq 2}\eta_{2i}) & \cdots & \cdots \\ \cdots & \cdots & \cdots & \cdots \\ -n_N\eta_{Ni} & \cdots & \cdots & n_N(\eta_N+\sum_{i\neq N}\eta_{iN}) \end{bmatrix} \cdot \begin{bmatrix}\frac{E_1}{n_1}\\ \vdots \\ \frac{E_N}{n_N}\end{bmatrix} \quad (5)$$

In the symmetric matrix system shown in Eq. 5, no information on shape modes and on natural frequencies is requested for SEA computation; this implies that the analysis could be considered well-conditioned only for relative high value of n_i, typically higher than 2-3 modes per band.

For lower n_i values, an hybrid approach is required; the information on natural modes and shapes are previously acquired from a preliminary FEM or BEM analysis and then the SEA computation can be carried out.

3.2 SUPERYACHT MODELLING

A predictive analysis of the noise propagation on board superyacht by using SEA requires the creation of a global model of the vessel, as shown in Fig. 2.

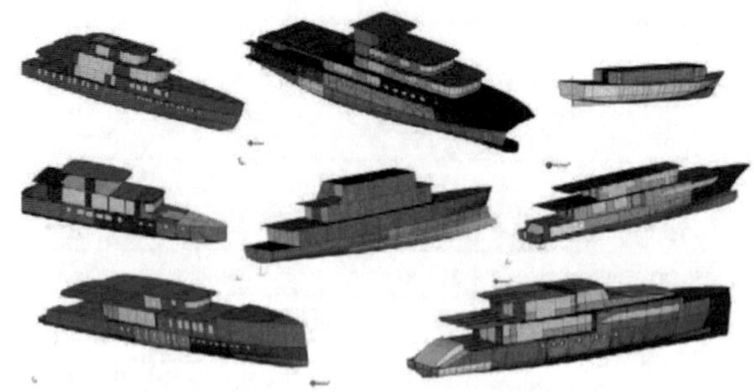

Figure 2: Examples of SEA global models for superyacht applications

The structural layout of the vessel (plates, longitudinal stiffeners and transversal frames) have to be created by using SEA plates or beams. This typology of elements, that could be actually more subdivided into different kind of plates (plane, curved ribbed), uses different transmission path for noise and vibration and implies also different coupling loss factor in the transmitted power Eq. 3.

While the creation of the structural layout could be considered as easier with respect to a complete FEM model due to the absence of small elements and nodes to be linked, the SEA procedure requires also the creation of internal subsystem, say "acoustic cavity", that are necessary for having the SPL as output.

Moreover, as shown in Fig. 3, internal subdivisions, as well as external plating, have to be carefully modelled following the vessel's insulation plans, in order to apply the correct noise control treatment (NCT) in each area.

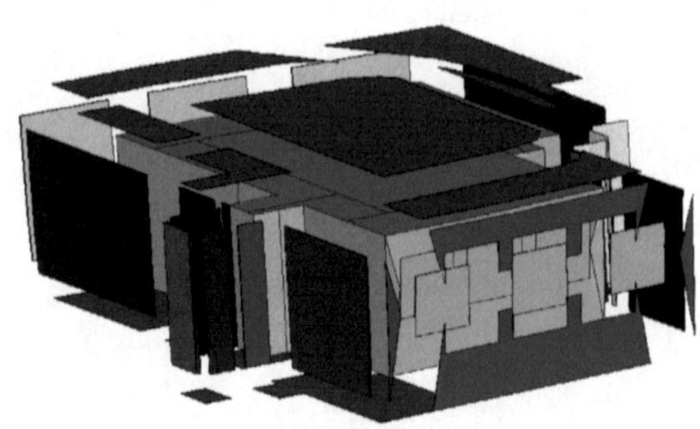

Figure 3: Internal panel subdivision of an owner's cabin

In green colour, the three windows have been clearly highlighted, because, as it will be discussed above, they cause the main power input contribution to the SPL of the cabin itself.

This means an increase in the pre-processing times and also implies a deep knowledge in the mechanical and dynamical (in terms of DLF and radiation efficiency) characterization of materials [7] [8].

3.3 TYPICAL RESULTS SCHEMES AND MAIN INPUT VARIABLES

As already stated, the main effort of SEA to naval architecture is the capability of predict the SPL in each zone of the vessel and the main energy path for SBN.

The overall SPL have to be compared with limit value imposed by IMO for what concerns charter yacht, and with Classification Societies rules for private yacht or if owner requires additional class notations.

In Fig. 4, a typical SPL diagram for an entire vessel is reported.

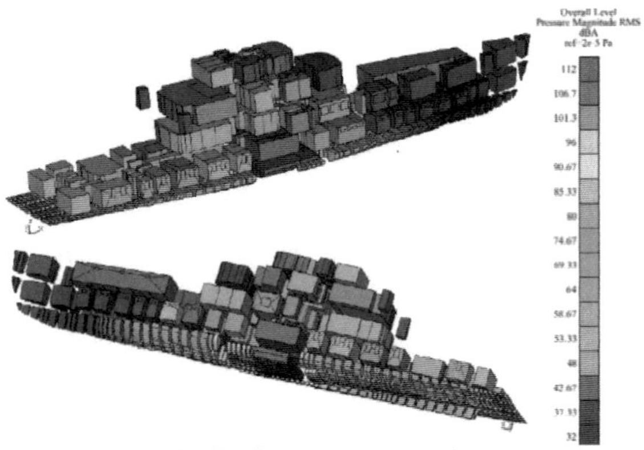

Figure 4: SPL distribution on a superyacht

Moreover, SEA analysis are able to plot (as shown in Fig. 5) the contribution to the total energy in each subsystem given by panels or adjacent cavity. In the following Fig. 5 the power input plot of the owner's cabin shown in Fig. 3 is reported.

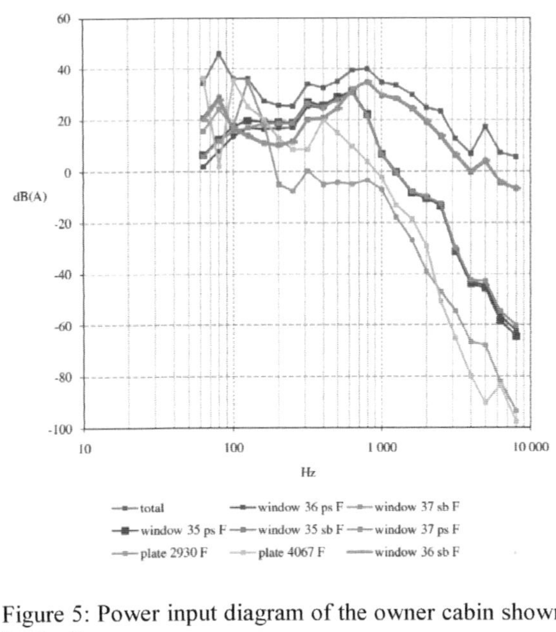

Figure 5: Power input diagram of the owner cabin shown in Fig. 3.

This tool is very helpful because, in case of rooms not complying with limits, it can clarify which panels radiates more noise and so where the transmission loss have to be improved.

In the aforementioned study case, the main contribution to the pressure level in the owner cabin is due to glazing windows in the mid frequency range; this behaviour is caused by the coincidence frequency (f_c) of the windows, which is at 800 Hz, causing the classical drop in the transmission loss, as shown in Fig. 6.

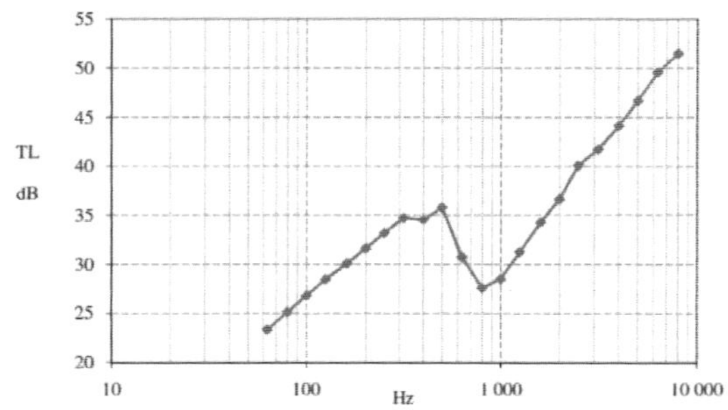

Figure 6: Transmission loss (TL) of owner's cabin windows

Looking in detail to Eq. 5, it appears clear that main input variables in the SEA procedure are the damping loss factor of each subsystem and the calculation of the power input. By the way, it is not sufficient to have these data in terms of magnitude, but they have to be quantified in the whole frequency range under consideration.

For these reasons, in next chapters of this work, numerical tools and experimental activities are presented in order to perform a reliable power input calibration (based on an hybrid approach) and to quantify the DLF of viscoelastic materials (that are largely used for engine foundations and as insulation materials) with easy and quick experimental activities

4. POWER INPUT CALIBRATION BY HYBRID SEA-FEM APPROACH

The SEA procedure requires a precise definition of the input power spectrum; for marine application, the principal sources are the propulsion engines, that are directly connected to the hull's structure thorough the engine foundation, as shown in Fig. 7.

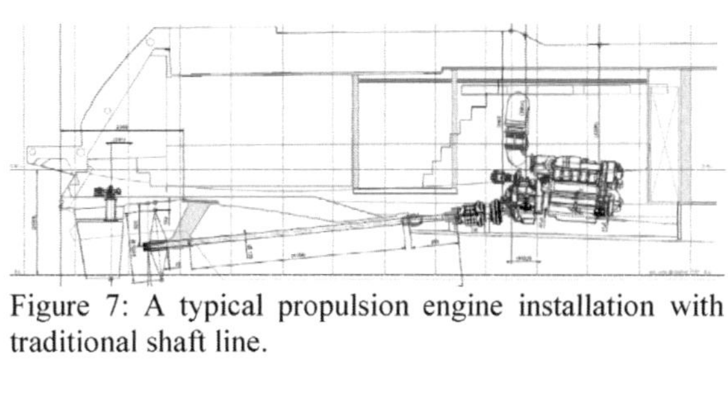

Figure 7: A typical propulsion engine installation with traditional shaft line.

The foundation, and in particular viscoelastic materials used for this application, are able to significantly modify the amount of energy transmitted to the hull's bottom and consequently the SBN propagation.

Therefore, the aim of this study is to test a methodology to obtain the power source to input in the foundations, starting from standard vibration data of main engine (usually given as vibration velocity spectrum).

The procedure herein presented can be ummarized by 4 main steps:

1. Creation of an hybrid model composed of FE model of structural foundation and SEA elements of the surrounding structures (see Fig 10);

2. Calibration of input forces (narrow band) to obtain the measured velocity spectra;

3. Calculation of the power input (1/3 octave band) representing the main engine action to inject in the full SEA global model;

4. Evaluation of the results in terms of sound pressure levels inside the vessel by means of SEA calculation with the computed power input.

(a)

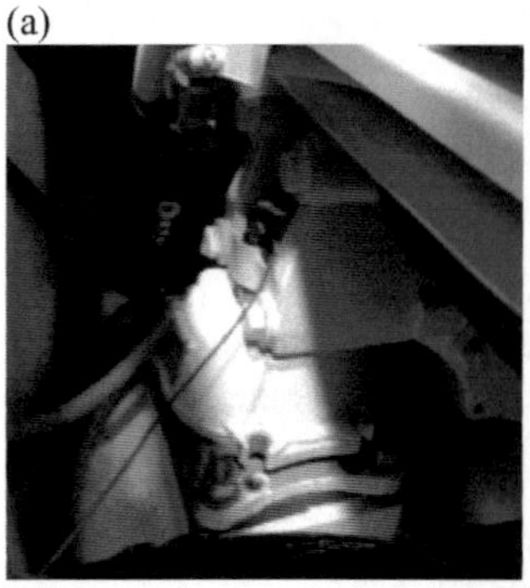

(b)

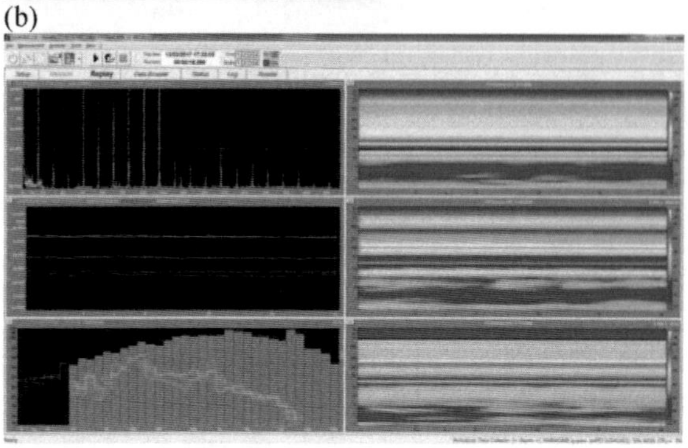

Figure 8: Velocity spectrum measurement, (a) accelerometers positioning and (b) response spectrum.

For this porpoise, two input data are required:

- Velocity spectrum measured on the main engine foundation below and above the resilient mounting (Fig. 8);

- FE model of structural foundation and SEA model of the whole vessel from which the hybrid model described at point 1 can be created.

The velocity spectrum has been measured by the instrumentation shown in Fig. 9; it consists in an impact hammer and piezoelectric accelerometers for the calculation of the frequency response function and a microphone for the calculation of the SPL.

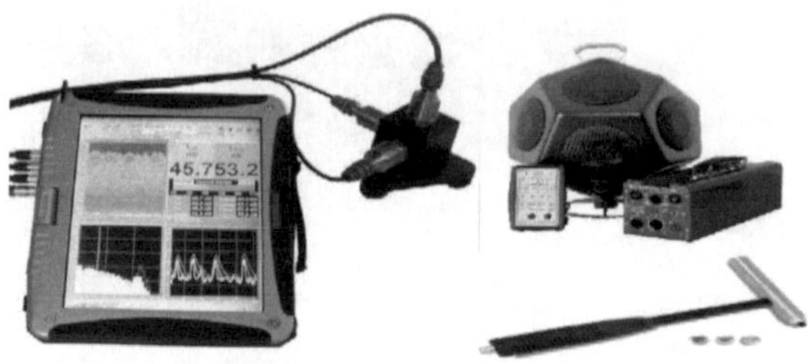

Figure 9: Instrumentation for N&V measurements on board.

The hybrid model of the engine foundation (Fig. 10) allows to perform a reliable calculation even for the low frequency range, where a full SEA approach could lead to higher uncertainties. Moreover, the same testing condition could be created in the coupled SEA-FE model, by using forces and velocity sensor in the same place of the experimental activity.

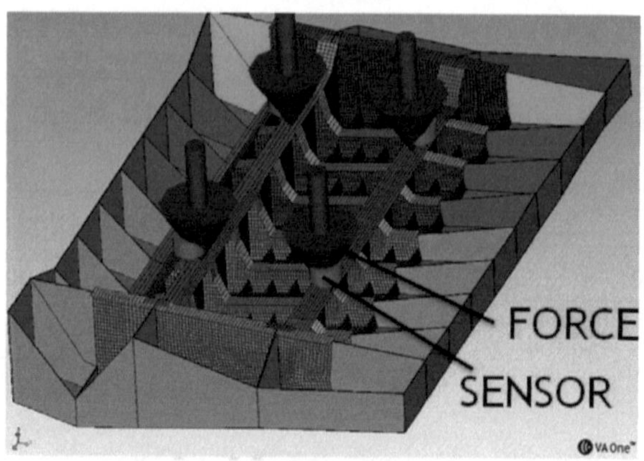

Figure 10: Hybrid model of the engine's foundation.

A local analysis of the hybrid model shown in Fig. 10 is carried out by applying a unit input force (F_{IN}). The velocity spectrum obtained by the numerical procedure (V_{IN}) have to be compared with the one obtained through out the experimental measurements ($V_{measured}$).

Being the FE analysis a linear static one [9], the superposition principle can be applied; so the correct input force (F_{INPUT}) can be calculated as:

$$F_{INPUT}(\omega) = F_{1N}(\omega) \cdot \frac{v_{measured}(\omega)}{v_{1N}(\omega)} \quad (6)$$

Once having known the total input force generated by the propulsion engines and the foundation's mobility (Y_z) (computed by the software), the power input spectrum can be easily calculated:

$$P_{INPUT}(\omega) = < F^2{}_{INPUT}(\omega) > \cdot Re(Y_z(\omega)) \quad (7)$$

The power input spectrum calculated by Eq. 5 can be applied to a global SEA model as shown in Fig. 11.

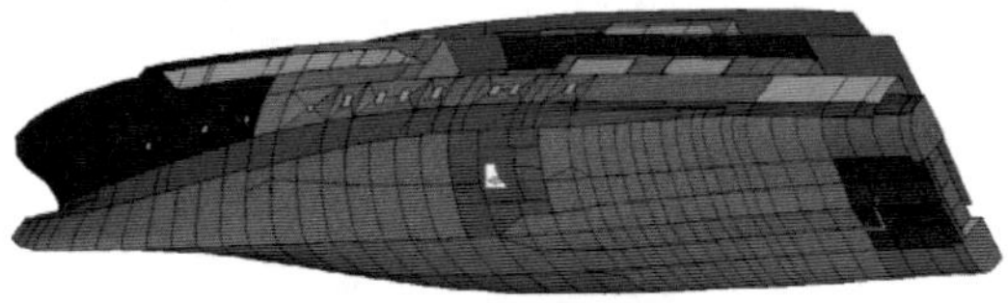

Figure 11: A global SEA model.

A comparison between the measured velocity spectrum and the velocity spectrum obtained from different numerical models, applying Eq. 6, is shown in the following Fig. 12.

As it can be easily noted FE model alone (i.e. not integrated in a hybrid model with SEA plates) is not reliable for the mid and high frequency range where the oscillations are caused by local resonances of the mesh grid that are not corresponding to the real phenomenon.

Using the abovementioned hybrid model the reliability of the results increases if compared to the measured velocity spectrum.

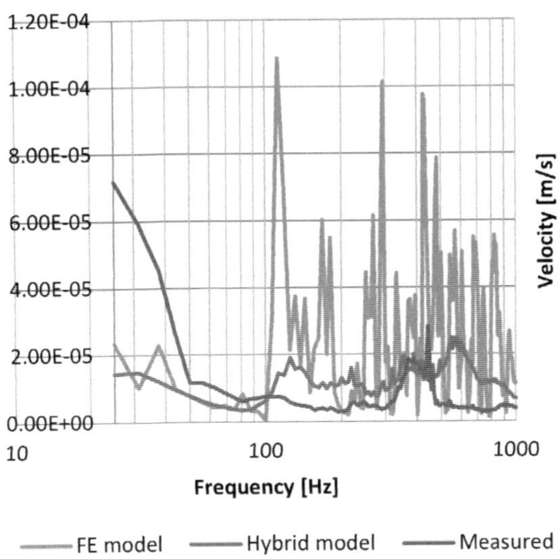

Figure 12: Velocity spectrum comparison

Fig. 13 shows a comparison between the SPL in the engine control room (adjacent to the engine room) measured onboard and the results of the SEA with the power input calculated with the described procedure.

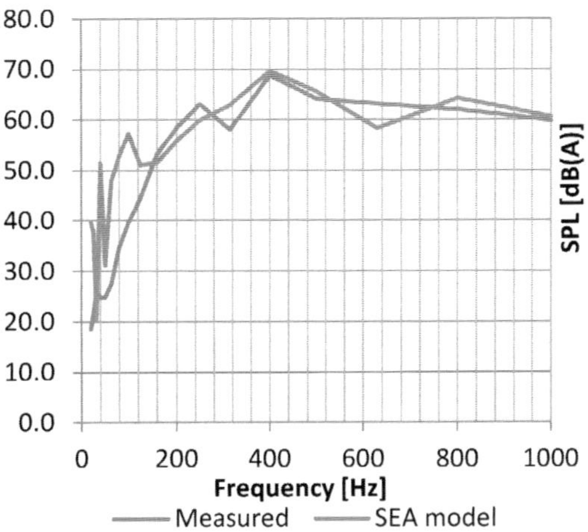

Figure 13: Comparison of the SPL in the engine control room

The general trend is comparable; there is a discrepancy at low frequency but a good correspondence around the peak frequency (400 Hz).

5. STRUCTURAL DAMPING CHARACTERIZATION

The second main uncertainty of the SEA calculation is the dynamical characterization of viscoelastic materials in terms of the damping loss factor.

The large use of this typology of material in superyacht firms is due to the relative low cost and weight and their high capability to insulate the noise propagation and to damp vibrations.

The characterization of DLF through the Oberst beam test requires very expensive machinery and the too small dimensions of specimens could affect the reliability of results.

Therefore, a RT60 based method have been developed for this porpoise.

5.1 EXPERIMENTAL EVALUATION OF THE DAMPING LOSS FACTOR

The experimental evaluation of the DLF have been studied on simplified steel specimens as reported in Fig. 14.

(a)

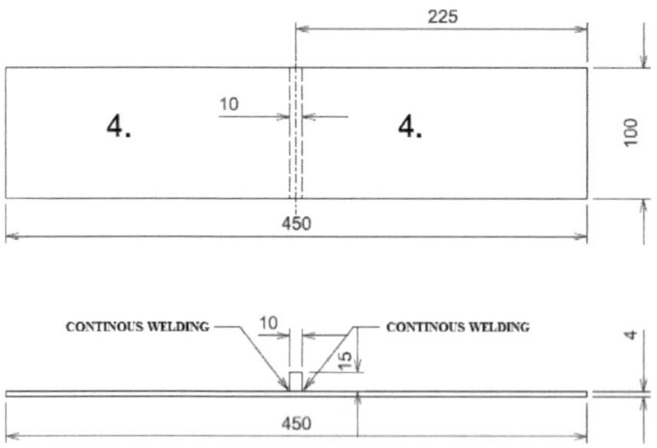

(b)

Figure 14: Specimen for DLF calculation, (a) dimensions and (b) set up.

The time history of the structural response has been studied by an impact test (Fig. 15) carried out with 10 different measures, in order to increase the reliability and reduce the deviation.

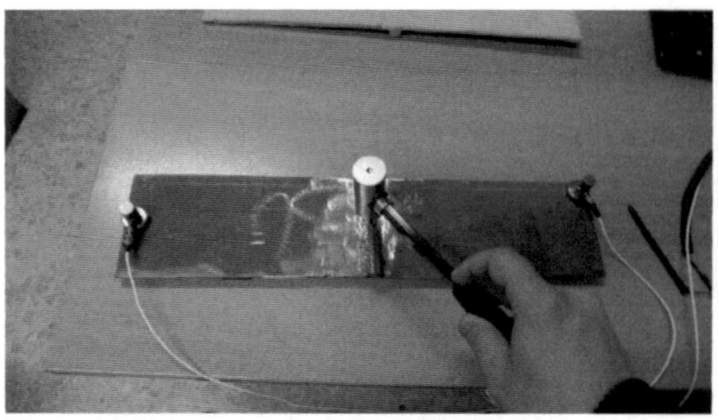

Figure 15: Impact test for DLF calculation

By analysis the time history and the sonogram of Fig. 16, it can be calculated the time that occur to reduce the vibration level of 60 dB (T_{60}) as it follows:

$$\alpha_{RT60} = \frac{\ln 10^6}{\omega T_{60}} = \frac{2.2}{f T_{60}} \tag{7}$$

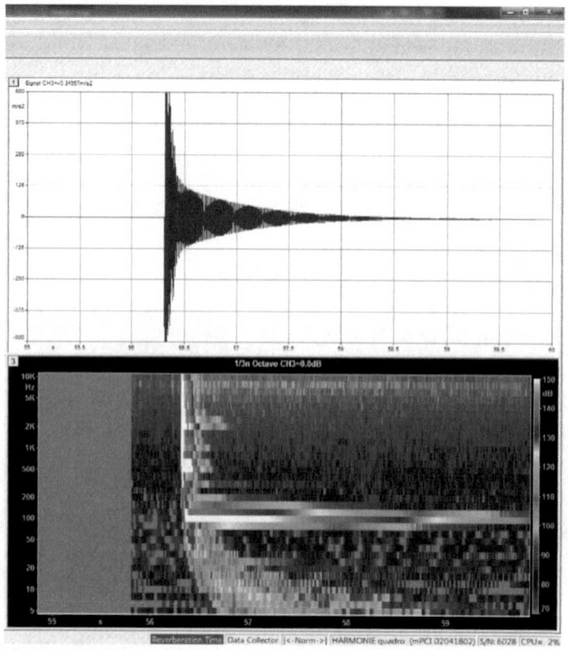

Figure 16: Time history and sonogram of the bare steel specimen

From the previous Fig. 16, it can be noted that the time decay is very long (more than 5 seconds) and that the 1/3 octave band that radiates more energy is at 100 Hz. It corresponds to the first flexural mode of the specimen, as reported in Fig. 17 for a FE analysis of the plates.

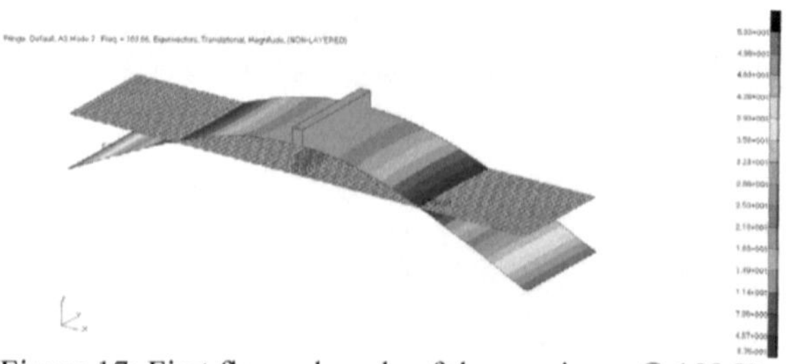

Figure 17: First flexural mode of the specimen @ 103 Hz

Once having calibrated and developed a standard procedure, that by the way is independent from the input source [8], several different viscoelastic materials have been tested, as shown in Fig. 18. They can be summarized in 3 different categories:

1. Viscoelastic materials constrained by a metallic plate;
2. Viscoelastic paints;
3. Adhesive meta-materials.

Figure 18: Specimen tested for the DLF estimation

5.2 DAMPING LOSS FACTOR INPUT FOR STATISTICAL ENERY ANALYIS

The results of the experimental activity is reported in terms of DLF spectra in Fig. 19.

For the very low frequency range, say under 100 Hz, the peak of DLF cannot be considered as fully reliable because the specimen has no natural modes and so the decay time is too short for a fully reliable measurement.

6. CONCLUSIONS

The more and more stricter rules and regulation in terms of vibration and noise level on board super and megayacht has shifted the focus point of naval architects to the N&V prediction and the insulation of vessels.

From this perspective, new numerical tools such as Statistical Energy Analysis have been developed in order to consider this tasks within the first project phases, where criticisms can be corrected with lower times and costs and in order to extend the reliable frequency range.

The level of uncertainties in a SEA calculation are higher then for a deterministic approach like FEA and they have to be carefully take into consideration while performing global analysis on vessels.
For these reasons, experimental activities have to be carried out to improve the confidence of analysis to an error up to 3 dB.

In this paper, a procedure for the calibration of the input power due to propulsion engines have been deeply studied by an hybrid SEA+FEA model. Moreover, also the effect of viscoelastic materials in terms of damping loss factor have been assessed.

This research is currently undergoing for what concerns the underwater radiated noise from superyachts.

7. REFERENCES

1. IMO Resolution MSC.337(91) "Adoption of the Code on Noise Levels on Board Ships" Annex 1 MSC 91/22/Add.1, 30 November 2012

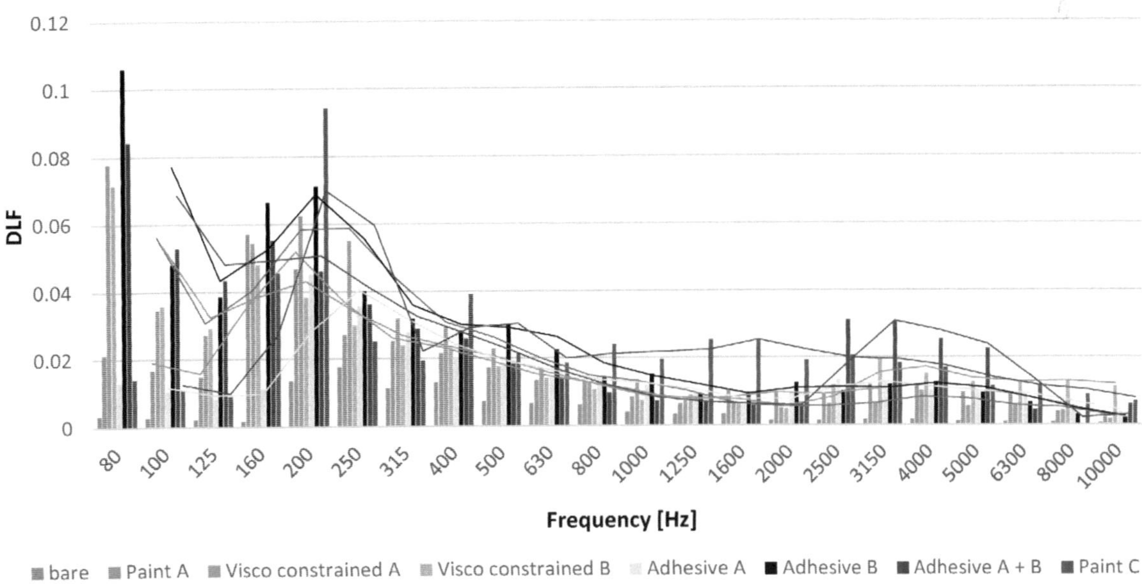

Figure 19: DLF spectrum for different viscoelastic specimens.

2. Rina Services http://www.rina.org/en.

3. Boote, D, Pais, T, Delle Piane, S.,(2013) "Vibration of superyacht structures", Proc 4th International Conference on Marine Structures, Finland, MARSTRUCT,

4. Pais, T, Boote, D, Kaeding, P. (2016). "Experimental and numerical analysis of absorber materials for steel decks". Proc 26th Int Offshore and Polar Eng Conf, Rodi, ISOPE

5. Lyon R.H., (1975) "Statistical energy Analysis of dynamic systems. Theory and Applications", The MIT Press.

6. Zhang J., Wang C., Sato T., (2013) "Advanced Markov Chain Monte Carlo Approach for Finite Element Calibration under Uncertainty", Computer aided civil and infrastructure engineering.

7. Gragnani L., Tonelli A., Boote D., Pais T., (2015)"Experimental investigation on the damping coefficient of laminated glass for large yachts", NAV 18th International Conference on Ships and Shipping Research, Lecco (IT)

8. Boote, D, Pais, T., Vergassola G., Tonelli A., Gragnani L., (2017) "On the damping coefficient of laminated glass for yacht industries", International Shipbuilding Progress.

9. MSC Software Corporation (2016), "Nastran User Guide", Newport Beach, CA, USA.

8. AUTHORS BIOGRAPHY

Prof. Dario Boote holds the current position of Ship Structure Professor at the Naval Architecture section of the Department of Electrical, Electronic, Telecommunications Engineering and Naval Architecture (DITEN) of the University of Genova. He is the Chairman of the Bachelor and Master Course in Yacht Design in La Spezia. His initial experiences include a long research activity in the field of Ship and Offshore Structures followed, since 2000, by an intense activity in the field of sailing and motor yachts. From 2009 to 2012 he has been Chairman of the V.8 ISSC Committee on "Yacht Design"

Mr. Gianmarco Vergassola holds the current position of PHD student at University of Genova. His research activity is devoted to the study of mechanical and dynamic characterization of glazing structures for naval and nautical architecture and to the noise propagation on super and megayacht.

Mr. Daniele Torre is currently part of the Technical Function, STRC (Structures and Calculation Tools) of RINA Services S.p.A., as structural engineer. Main expertise was gathered in the fields of noise and vibration on board, through several activities and studies for ships and motor yachts.

Mr. Angelo Tonelli in 1994 entered in the Research and Development (R & D) Department of Registro Italiano Navale in follow became RINA Services. Since then he has been working on different numerical and experimental activities related to merchant and naval ships in the fields of computational fluid-dynamics, risk analyses, finite element analyses and noise and vibration harshness (NVH) with statistical energy analyses.

THE INFLUENCE OF ADDED MASS IN SUPERYACHT VIBRATION ANALYSIS

T. Pais, **D. Boote** and **G. Vergassola,** DITEN, Università di Genova, Italy

SUMMARY

The prediction of the comfort level of superyachts in the early design stage, based on suitable methodologies to predict the structural dynamic behaviour, is the present challenge for designers. In case the resulting noise and vibration levels were not acceptable, the shipyard can apply corrective actions to reduce their effects before the construction phase. The usual way to perform a reliable predictive analysis is by using finite element technique; in this perspective the result reliability depends on the numerical model quality. In case of a ship an important parameter which heavily influences the dynamic behaviour of a numerical model is the water added mass; in this paper, this subject is studied by theoretical and numerical methods. A comparison between the added mass obtained by the well known Lewis formulation and the numerical results is carried out. In a first phase, a simple study case is investigated and after that the added mass of a study case superyacht has been calculated by a numerical approach.

1. INTRODUCTION

Yacht designers and builders are continuously looking for new solutions to reduce construction costs and to improve the quality and innovation of their vessels. In the case of superyachts over 30 meters in length, performances are no longer a primary goal and the efforts of technical offices are mostly addressed on other aspects related to the aesthetic impact of the project and to the on board comfort [1]. From this point of view, vibrations and noise represent a difficult issues to deal with for designers, both in the initial phase of the project, when it is necessary to have preliminary information about the response of the structure not yet defined, and during construction, in case some critical behaviours arose in any part of the structure [2]. Given the objective difficulty in making any change to the dynamic behavior of hull structure after construction, it is extremely important to perform FEM predictive analyses to identify the natural frequency of the hull and of local structures, such as decks and bulkheads, and then their response to exciting loads induced by propellers, engines and waves. Vibration problems, in particular, are certainly more critical in the case of steel and light alloy yachts due to the low damping, even if it is well known that also FRP vessels are not free from this kind of problem.

In addition, ship vibrations are greatly influenced by the inertial forces of surrounding fluid. These forces are proportional to the acceleration of wetted surface, and can consequently be represented by fictive fluid mass that vibrates along with the ship hull. This fictive mass is called "hydrodynamic added mass" which, in addition to the ship structure mass, represents the "virtual mass" [3].

The first phase of this activity is represented by a comparison of the added mass obtained by the Lewis formulas and by the numerical model of a steel cylinder of 25 m in length. Then the attention is focused on the more complex structure of a super yacht.

2. SHIP VIBRATION STUDY

Ship hulls are elastic structures in which dynamic loads generated by propellers and by propulsion and maneuvering machineries on board, may induce vibrational problems [4, 5]. Many studies are carried out on this subject since the early decades of the twentieth century, and several empirical-statistical approaches are developed, just as some examples, by Todd [6], Burrill [7] and Lewis [8].

Since then, considerable advances are reached in the theoretical and experimental assessment of ship vibration, both on predictive analysis methods and measurement methodologies. On the other hand, the solution of vibration problems on already built ships still represent a challenging task.

In recent years, the definition of the dynamic behavior of a new vessel, has become a standard practice. The builder in this way assumes the task that vibration limits should not be exceeded and, if this does not happen, to identify and apply proper reliable solutions. In the preliminary design phase and during the subsequent construction phases, the yard performs a predictive study of the vibration levels of the structure or it can delegate this task to an independent technical office. The flow diagram in Figure 1 can synthesize the overall procedure [9].

Preliminary vibration analyses should be already started since the concept design; by this approach, if some vibration problems are already detected, they can be analyzed and solved with minor impact on the subsequent design phases with considerable benefit in terms of costs and time.

Nevertheless, in this phase of the project, a detailed study of local vibrations is not still possible because the structural layout of reinforced and secondary reinforcements is not yet developed. Generally, this initial analysis focuses on areas that are deemed critical on the base of the designer experience. The vibration behavior of a vessel can be classified from two different points of view: the ship area where they take place and the vessel-operating regime at which vibrations occurs.

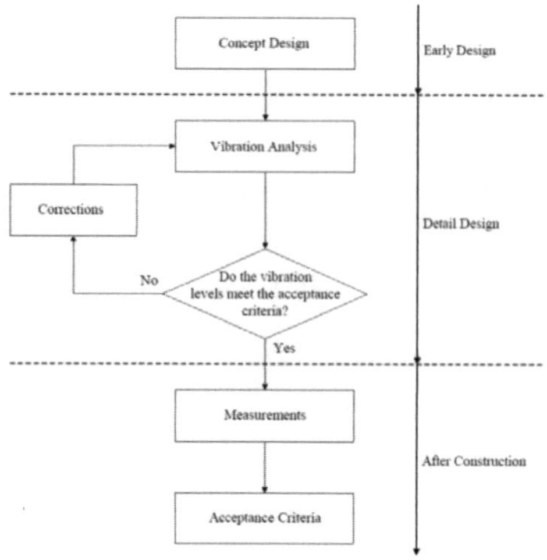

Figure 1: Overall Procedure for Ship Vibration Assessment [9]

Passing to the more advanced design phase, the "Detail design", vibration analysis can be approached by theoretical methods on simplified models, by semi empirical methods based on available statistical data or by numerical analyses carried out by proper finite element codes. In this phase, structural drawings of the whole vessel and the corresponding load conditions are now necessary and sufficiently detailed [9].

By this approach, critical dynamic situations can be detected with precision and easily overcome. When the vessel will be completed, the vibrational behavior should be verified by appropriate real scale measurements, usually carried out during sea trials before the ship delivery.

3. ADDED MASS

Due to the high density of the fluid in which they operate, ships are subjected to considerable inertial effects. When moving, the ship drags with her a significant portion of water, the mass of which can be considered associated to the ship mass: this water mass is called "Added mass" [10].

The hydrodynamic added mass is usually of considerable entity; as an example, in the case of a sphere immersed in a fluid, moving perpendicularly to the free surface, the added mass is equal to half the mass displaced by the sphere itself. Consequently, the presence of the added mass cannot be neglected in a ship vibration analysis.

Burrill, in the 30ies of last century, elaborated an initial empirical formulation of the first vertical mode of vibration of a ship [7].

$$N\nu = \varphi \cdot \sqrt{\frac{I}{(\Delta+\Delta_1)\cdot L^3\cdot(1+Ns)}} \qquad (1)$$

where:

Nv frequency of the first mode of vertical vibration, expressed in cycles per minute
I momentum of inertia of the main section, expressed in feet^4;
Δ ship displacement, expressed in tons;
$\Delta 1$ fluid added mass;
L ship length between perpendiculars;
Ns Lockwood-Taylor shear correction factor;
φ empiric constant.

As it is possible to note in the above equation, the added mass plays a decisive role in the determination of natural frequencies of a ship. For the determination of the term $\Delta 1$, different methodologies are available. The Lewis method, still remains the most used one [8].

4. ADDED MASS CALCULATION, LEWIS METHOD

For the vertical plane motions, the added mass associated to the numerical model is manually calculated by using the Lewis formulation. This methodology involves the identification of different "sectional" hydrodynamic masses (thus masses per unit of length) which, integrated over the length of the submerged body, provide the total added mass. In order to identify the sectional added mass, Lewis provides a series of curves that, depending on the shape of a generic section and from the quantity $\frac{B(x)/2}{T(x)}$, they return a "Lewis coefficient".

After the total added mass is calculated, it must be finally multiplied by a corrective coefficient J, introduced to take into account the three-dimensional effects acting on the body. Hereinafter, formulas for calculating the mass addition are illustrated.

Given a ship section, the following dimensions are defined as shown in Fig. 2:

- d(x) the draught;
- b(x) the half breadth of the ship section at waterline;
- S(x) the area of the immersed part of the shipframe;
- p is the centroid of the immersed part of the ship section area;
- c is the position of the center of torsional oscillations.

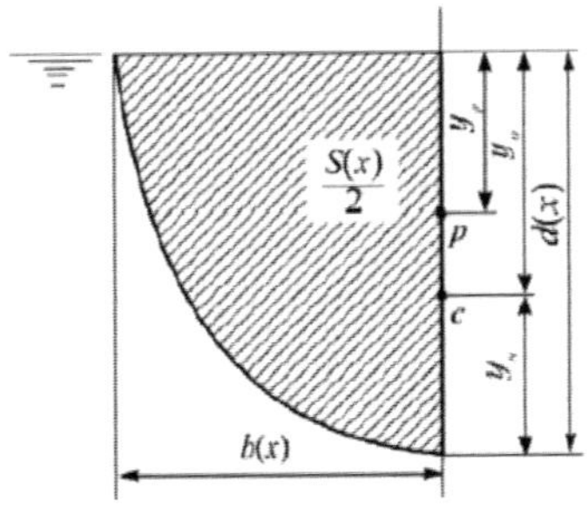

Figure 2: Scheme of (half of) the immersed part of a shipframe [10]

The following entities are introduced:

$$\sigma(x) = \frac{(x)}{2\, b(x) d(x)} \qquad q = \frac{d(x)}{b(x)} \tag{2}$$

and the following auxiliary coefficient defined:

$$a = \frac{1}{2}\left[3(1+q) - \sqrt{1 + 10q + q^2 - \frac{32\sigma q}{\pi}}\right] \tag{3}$$

$$c_v = 1 + (1 + q - a) \cdot (q - a) \tag{4}$$

c_v is the section two-dimensional added mass coefficient, also known as the so called "Lewis form coefficient".
The added mass of ship sections in the vertical direction can be found by the formula:

$$\lambda_{33} = J_n c_v \left(\frac{\pi}{2} \rho b^2\right) \tag{5}$$

The index n in Eq. 5 indicates that added mass is dependent on the vibration mode. In this case, the reduction J factor is defined as:

$$J_n = 1.02 - 3\left(1.2 - \frac{1}{n}\right)\frac{B}{L} \tag{6}$$

where B is the water line breadth amidships and L is the ship length. The formula only applies to vertical vibrations with n = 2-5.

5. NUMERICAL COMPUTATION OF ADDED MASS

Numerical methods for added mass calculation are nowadays available and implemented in commercial FE packages. In the case of the Nastran software, the boundary element method is considered.

Nastran is provided with a special function which allows to take into account the possible presence of fluids internal or/and external to the model. The command is called "MFLUID" and the relative theory is taken from the Nastran theoretical manual [11].

The Helmholtz method used by MSC Nastran solves Laplace's Equation by distributing a set of sources over the outer boundary, each producing a simple solution to the differential equation. By matching the assumed known boundary motions to the effective motion caused by the sources, we can solve a linear matrix equation for the magnitude of the sources. The values of the sources determine the effective pressures and, thereby, the forces on the grid points. Combining all of these steps into a matrix equation results in a virtual mass matrix as derived below.

If σ_j is the value of a point source of fluid (units are volume flow rate per area) located at position r_j, and assumed acting over an area A_j, the vector velocity $\dot{u}_i$ at any other point r_i is:

$$\dot{u}_i = \sum_j \int_{A_j} \frac{\sigma_j \sigma_{ij}}{|r_i - r_j|^2} dA_j \tag{7}$$

Note that the gradient of the vector $\dot{u}_i$ is the potential function which satisfies Laplace's Equation on a term by term basis.

The other set of necessary equations are the pressures, p_i, at any point i, in terms of the density ρ, sources and geometry, namely

$$p_i = \sum_j \int_{A_j} \frac{\rho \dot{\sigma}_j e_{ij}}{|r_i - r_j|} dA_j \tag{8}$$

where e_{ij} is the unit vector in the direction from point j to point i. The results of integrating Eqs. 7 and 8 over the finite element surfaces are collected respectively in two matrices:

$$\{\dot{u}\} = [\chi][\sigma] \tag{9}$$

And

$$\{F\} = [\Lambda]\{\dot{\sigma}\} \tag{10}$$

where F are the forces at the grid points. The matrix $[\Lambda]$ is obtained by integrating Eq. 8. An additional area integration is necessary to convert the pressures to forces. A mass matrix may now be defined using Eqs. 9 and 10 as:

$$\{F\} = [M^f]\{\ddot{u}\} \tag{11}$$

where the virtual fluid mass matrix, [Mf] is

$$[M^f] = [\Lambda][\chi]^{-1} \tag{12}$$

6. COMPARISON OF ANALYTICAL NUMERICAL METHOD FOR A CYLINDER

As first study case, a FE model of a steel cylinder of 25 m in length, 1 m diameter and 15 mm of thickness is created. For this numerical model, only shell elements are used.

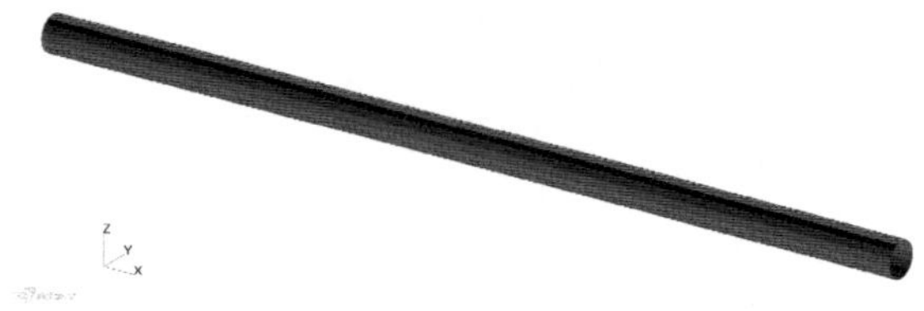

Figure 3: Finite element of a steel cylinder

Before the added mass calculation, the natural frequencies of the cylinder are calculated. The Euler and Timoshenko equations are used to calculate the natural frequencies of cylinder free at both ends. The Timoshenko equation differs from the Eulero one because it takes into account the "shear factor" k [12].

Eulero equation: $$\omega_j = \frac{(j+0.5)^2\pi^2}{L^2}\sqrt{\frac{EI}{\rho A}} \qquad (13)$$

Timoshenko equation:

$$\omega_j = \frac{(j+0.5)^2\pi^2}{L^2}\sqrt{\frac{EI}{\rho A}}\sqrt{\frac{1}{1+\frac{(j+0.5)^2\pi^2 I}{L^2 A}\left(1+\frac{E}{kG}\right)}} \qquad (14)$$

In Table 1 the comparison between results obtained by Eqs. 13 - 14 with the FE model are shown. The difference between natural frequencies calculated by Timoshenko equation and those obtained by FEM is very low.

Table 1: Comparison of natural frequencies by different methodologies

Mode shape	Eulero [Hz]	Timoshenko [Hz]	FEM [Hz]	Error
1	10.31	10.17	10.21	0.37%
2	28.39	27.39	27.59	0.72%
3	55.64	51.99	52.57	1.13%
4	91.97	82.58	83.90	1.60%
5	137.39	117.85	120.38	2.15%
6	191.90	156.59	160.87	2.73%

After the estimation of the natural frequencies relative to the mode shapes of the cylinder, the added mass is calculated. The cylinder is immersed in seawater for half the diameter. For calculations, the length of the model is divided into twenty sections with constant spacing, as shown in table 2.

Table 2: Added mass of the cylinder

Section	X [m]	b [m]	T [m]	Area [m²]	c_v	$c_v\left(\frac{\pi}{2}\rho b^2\right)$
1	0	1	0.5	0.393	1.0	402.517
2	1.25	1	0.5	0.393	1.0	402.517
3	2.5	1	0.5	0.393	1.0	402.517
4	3.75	1	0.5	0.393	1.0	402.517
5	5	1	0.5	0.393	1.0	402.517
6	6.25	1	0.5	0.393	1.0	402.517
7	7.5	1	0.5	0.393	1.0	402.517
8	8.75	1	0.5	0.393	1.0	402.517
9	10	1	0.5	0.393	1.0	402.517
10	11.25	1	0.5	0.393	1.0	402.517
11	12.5	1	0.5	0.393	1.0	402.517
12	13.75	1	0.5	0.393	1.0	402.517
14	16.25	1	0.5	0.393	1.0	402.517
15	17.5	1	0.5	0.393	1.0	402.517
16	18.75	1	0.5	0.393	1.0	402.517
17	20	1	0.5	0.393	1.0	402.517
18	21.25	1	0.5	0.393	1.0	402.517
19	22.5	1	0.5	0.393	1.0	402.517
20	25	1	0.5	0.393	1.0	402.517
					TOT	**10.06 [tones]**

Table 3: Added mass value for each mode shapes

Mode shape	J_n	λ_{33} [Kg]
1	0.9116	9173.188
2	0.9393	9452.099
3	0.7872	7921.253
4	0.8160	8211.714
5	0.7078	7122.069
6	0.7350	7396.722

In this case the correction factor J is:

$$J_n = \frac{16}{\pi^2}\sum_m^\infty\left(\frac{m}{m^2-n^2}\right)\left[1+\frac{R\pi m}{L}\frac{K_0(R\pi m/L)}{K_1(R\pi m/L)}\right]^{-1} \qquad (15)$$

where K_0 and K_1 are modified Bessel functions of the 2nd kind: 0th and 1st order respectively, while m = 1,3,5,... if number of vibration nodes n is even, else m = 2,4,6,...
In Table 3, the added mass value for each mode shape is shown.

The only way to verify the added mass in the FE calculations is to compare the dry and wet natural frequencies. The displacement of cylinder is 9130 kg; by the following equation, the added mass is obtained [10].

$$\lambda_{33,n} = \Delta\left[\left(\frac{\omega_{n,dry}}{\omega_{n,wet}}\right)^2 - 1\right] \tag{16}$$

In Table 4 the comparison between the added mass obtained by FEM and by the theory is shown.

Table 4: Comparison of added mass values

Mode shape	ω_{dry} [rad/s]	ω_{wet} [rad/s]	λ_{33} FEM [Kg]	λ_{33} Theoretical [Kg]	Error
1	64.15	45.55	8976.49	9173.19	2.14%
2	173.33	122.15	9252.77	9452.10	2.11%
3	330.33	267.08	4836.64	7921.25	38.94%
4	527.18	404.79	6355.57	8211.71	22.60%
5	756.37	571.28	6874.48	7122.07	3.48%
6	1010.78	791.87	5745.55	7396.72	22.32%

7. SUPERYACHT NUMERICAL MODEL

The study case vessel assumed for the investigation is a 54 meters superyacht buil by a famous Italian shipyard, and kindly made available for the present study.

The FE modeling of ship structures is the most laborious and time consuming part of a dynamic analysis. As well known the modeling of a global structure by a FE code is necessarily subjected to some basic approximations in order to reach a sufficiently good compromise between the model accuracy and the calculation complexity in terms of cost and time. In other word, a perfect reproduction of a ship structure by FE software could be counterproductive in practical terms, as well as impossible for technological limitations.

As the objective of this study is that to perform global and local analyses all structural components have been accurately reproduced. The mesh is created using "SHELL" elements for plating and main reinforcements such as keelsons, floors and girders. For secondary stiffeners simple "BEAM" elements have been preferred to keep the model dimension as low as possible.

The hull geometry and structure lay out is imported from a 3D model previously created by a rendering software, as shown in Figure 4.

Figure 4: 3D render model

In Figure 5, the numerical model of the yacht, as realized by Patran Software, is shown.

Figure 5: Patran numerical model

In Figures 6 and 7 some particular details of the model are shown.

Figure 6: Numerical model of Sun Deck

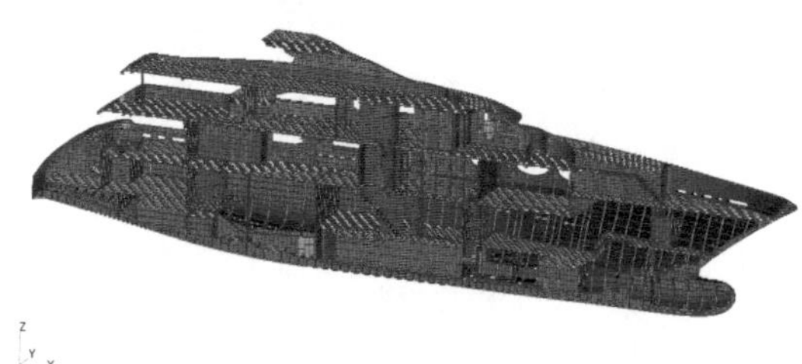

Figure 7: Inner view of the numerical model

In total, the numerical model of the yacht consists of 119'907 elements, 72'916 nodes and 396'144 DOFs.
For the attribution of the nonstructural masses to the numerical model, two methods are available in Patran software: distributed or lumped masses [11]. All weights less than 1 Kg/m^2 are applied as distributed load, such as insulation, flooring, marble, pipes, cables, filler, painting, furniture etc. In Figure 8 the distributed mass are shown.

In case of more concentrated weights they are applied to the numerical model as lumped masses on points where the weight insists on. In Figure 9, as an example, the gearbox and electric generator masses are shown in the engine room.

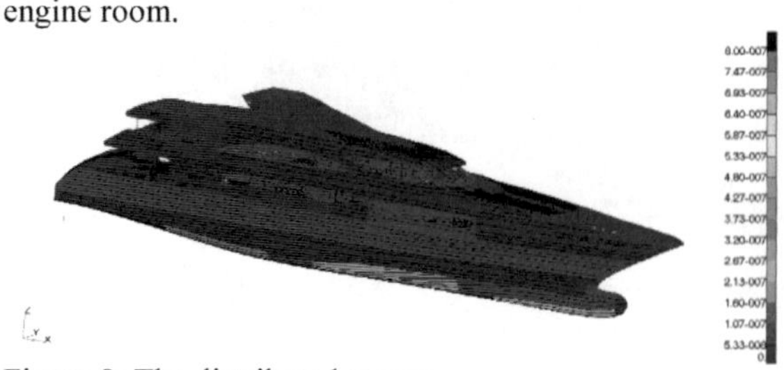

Figure 8: The distributed masses

Figure 9: Examples of lumped masses application

8. ADDED MASS FOR A SUPERYACHT

I position	II position	III position
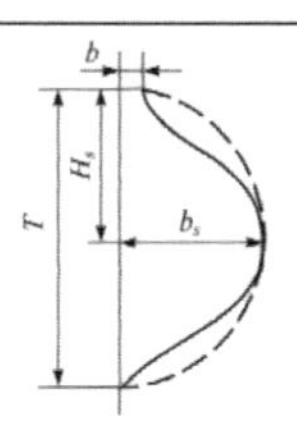	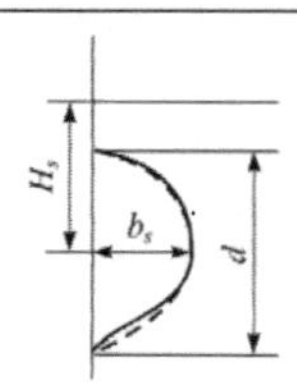	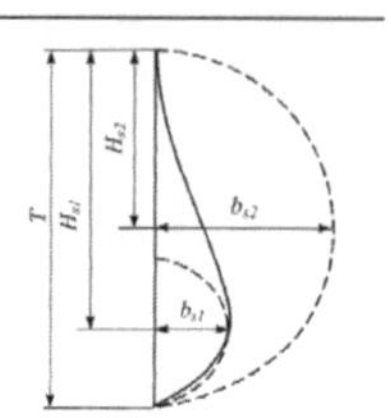
$\lambda_{33} = 0.5\rho\pi b_s^2$	$\lambda_{33} = \rho\pi b_s\left(1 - \frac{b_s^2}{2H_s^2}\right)$	$\lambda_{33} = \rho\pi b_{s1}^2\left(1 - \frac{b_{s1}^2}{2H_{s1}^2}\right)$

Figure 10: Added mass for bulb-type ship sections (54).

After the simple example of the cylinder, the attention has been focused on the more complex structure of a super yacht.

The length of the yacht is dived into twenty equidistant sections by the Lewis procedure. For the typical positions of bulb-type ship sections shown in the Figure 10, the added mass is obtained by the following formulas.

In Table 5, the added mass for each 2D sections of the yacht is calculated and in Figure 11 it is reported in the graph.

Table 5: Added mass of the yacht

Section	x [m]	b [m]	T[m]	Area [m^2]	c_v	$c_v\left(\frac{\pi}{2}\rho b^2\right)$
1	0	8.947	1.117	8.114	1.014	32669.301
2	2.5	8.947	1.117	8.114	0.935	30125.092
3	5	8.957	1.505	10.326	0.903	29173.454
4	7.5	8.972	1.927	13.177	0.891	28859.868
5	10	9.452	2.316	16.175	0.874	30000.000
6	12.5	9.011	2.623	18.936	0.904	29531.295
7	15	9.033	2.831	21.240	0.923	30325.893
8	17.5	9.056	2.951	22.833	0.942	31109.275
9	20	9.042	2.989	23.483	0.955	31427.816
10	22.5	8.940	2.993	23.246	0.955	30733.725
11	25	8.706	3.000	22.170	0.939	28646.261
12	27.5	8.295	3.000	20.264	0.912	25256.801
13	30	7.765	3.000	17.666	0.872	21161.143
14	32.5	6.799	3.000	14.709	0.847	15756.239
15	35	5.734	3.000	11.641	0.820	10846.932
16	37.5	4.546	3.000	5.938	0.750	6238.904
17	40	3.326	3.000	5.940	0.787	3502.099
18	42.5	0.714	3.000			820.805
19	45	0.676	3.000			736.811
20	47.5	0.671	1.742			195.671
					TOT.	**1005 [tones]**

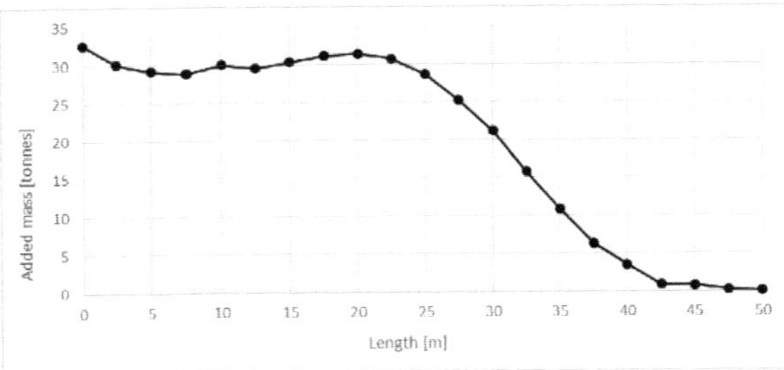

Figure 11: Length distribution of the added mass

The sectional 2-D added mass is multiplied by the correction factor defined in Eq. 6.

Table 6: Added mass value for each mode shapes

Mode shape	J_n	λ_{33} [tonnes]
1	0.67	673.68
2	0.59	589.89
3	0.55	547.99
4	0.52	522.86

The modal analysis is performed before and after the introduction of the added mass. In Figure 12, the first mode at 7.32 Hz of the dry structure is shown. In Figure 13, the first mode at 4.89 Hz with wet structure is shown.

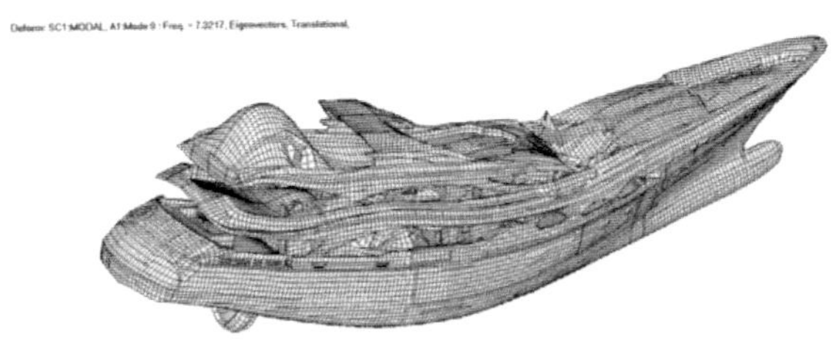

Figure 12: First mode of the dry structure

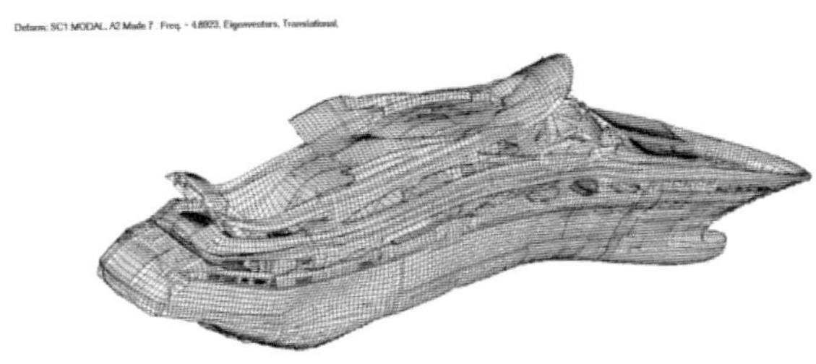

Figure 13: First mode with wet structure

From the numerical model, the added mass is obtained comparing the dry and wet natural pulsations by Eq 16.

Table 7: Comparison of the added mass value

Mode shape	ω_{dry} [rad/s]	ω_{wet} [rad/s]	λ_{33} FEM [tonnes]	λ_{33} Theoretical [tonnes]	Error
1	45.99	30.73	684.3	673.68	1.55

9. CONCLUSIONS

The expectations of owners and the Classification Societies requirements about superyacht onboard comfort impose to designers to sensitize more and more to the problem of predicting vibroacoustic behaviour of ship structures. In addition to comfort aspects, the dynamic behaviour of hull structures heavily influences the efficiency, safety and health of passengers and crew as well. In order to correctly evaluate the influence that all parameters have on the dynamic behaviour of any structure, the determination of the vibration, noise, and their routes of transmission results to be necessary.
In this perspective the only way to assess this subject is through the extensive use of FE technique. In this case the quality of results depends on the finite element model accuracy and on the exact evaluation of the associated added mass. The aim of the present research is just to deepen the knowledge of the added mass value, a parameter which has a great influence on the dynamic behaviour of a ship.

In the first part of this study, a comparison between the added mass obtained by the well-known Lewis formulation and the numerical results on a submerged cylinder has been carried out. The resulting error between the two approaches has been found to be very small for the low frequency range shown in Table 4.
In the second phase, the added mass of a superyacht is investigated. The resulting error relatively to the first mode shape between the theoretical added mass and the value got from the numerical model is again very small as shown in Table 7. In this phase, only the first natural frequency of the vessel has been taken into account; as a matter of fact the huge and complicate structure of a superyacht makes very difficult to individuate two consecutive mode shapes and, as a consequence, to evaluate the existing frequency differences.

10. REFERENCES

1. D.Boote, T. Pais, S. Dellepiane, "*Vibrations of Superyacht Structures: Comfort Rules and Predictive Calculations*", Proceedings of Marstruct 2013, Analisys and Design of Marine Structures, 4th International Conference on Marine Structures, Espoo, Finland, March 2013.

2. D. Boote, T. Pais, E. Camporese, "*Numerical and Experimental Analysis of the Dynamic Behaviour of Large Yacht Superstructures*", Proceedings of "Design and Construction of Super and Mega Yachts" Conference, Royal Institution of Naval Architects, Genova, May 2013.

3. FALTINSEN, O.M.: "*Sea Loads on Ships and Offshore Structures*", Cambridge Ocean Technology Series, 1990.

4. M. Biot, D. Boote, E. Brocco, L. Moro, T. Pais, S. Delle Piane, "*Numerical and experimental analysis of the dynamic behavior of main engine foundations*", International Ocean and Polar Engineering Conference ISOPE 2015, Hawaii, USA, June 2015.

5. M. Biot, D. Boote, E. Brocco, L. Moro, P.N. Mendoza Vassallo, T. Pais, "*Validation of a Design Method for the Simulation of the Mechanical Mobility of Marine Diesel Engine Seatings*", Proceedings of Transport Means 2014 Conference, Kaunas University of Technology, Lithuania, 23rd – 24st October 2014.

6. F. H. Todd, "Ship hull vibration", Edward Arnold & CO, London, 1961.

7. L. C. Burrill, "Ship Vibration; Simple Methods of Estimating Critical Frequencies", Transactions NECI, Vol. 51, 1934.

8. F. M. Lewis, "The inertia of the water surrounding in a vibrating ship", Trans. SNAME 37,1–20, 1929.

9. American Bureau of Shipping, "Guidance Notes on Ship Vibration", January 2015.

10. A. I. Korotkin, "Added Masses of Ship Structures", Springer, 2007.

11. Nastran documentation, "Dynamic analysis User's Guide", MSC Software, 2013.

12. Y. Yin, D. Zhao, H. Cui, M. Hong, "*Prediction method of natural frequency for ship's overall vertical vibration*", Brodogradnja/Shipbuilding Volume 65 Number 3, 2014.

11. AUTHORS BIOGRAPHY

Tatiana Pais holds the current position of research assistant in the Naval Architecture section of the Department of Electrical, Electronic, Telecommunications Engineering and Naval Architecture (DITEN) of the University of Genova. Her research activity deals with the dynamic behavior of yacht structures.

Dario Boote holds the current position of Ship Structure Professor at the Naval Architecture section of the Department of Electrical, Electronic, Telecommunications Engineering and Naval Architecture (DITEN) of the University of Genova. He is the Chairman of the Bachelor and Master Course in Yacht Design in La Spezia. His initial experiences include a long research activity in the field of Ship and Offshore Structures followed, since 2000, by an intense activity in the field of sailing and motor yachts. From 2009 to 2012 he has been Chairman of the V.8 ISSC Committee on "Yacht Design".

Mr. Gianmarco Vergassola holds the current position of PHD student at University of Genova. His research activity is devoted to the study of mechanical and dynamic characterization of glazing structures for naval and nautical architecture and to the noise propagation on super and megayacht.

LNG AS A YACHT FUEL

E R Harland, Houlder Ltd, UK, J D Strachan Houlder Ltd, UK

SUMMARY

The use of LNG provides a clean and environmentally friendly powering option that is being adopted by a number of Cruise and Passenger Ships. This paper covers the potential for using LNG as a Yacht fuel.

The paper includes an explanation of the additional systems required to ensure safe operation and compliance with the latest international legislation.

LNG is an effective alternative fuel but with current supply infrastructure should be considered as part of a dual fuel solution. LNG would be especially advantageous for providing clean power and reducing HVAC power consumption.

1. INTRODUCTION

This paper draws on Houlder's recent experience with LNG as a ship fuel (Fergusson NB801/2). The lessons and design challenges of using LNG as a fuel have been reviewed in the context of the specific demands of the yacht sector and the option of LNG as a Yacht fuel investigated.

LNG is the available, clean and the most environmentally friendly fossil fuel. LNG is being adopted and the fuel of choice by many owners on newbuild Cruise and Passenger Ferry projects keen to take advantage of its environmental benefits and the large natural gas reserves in the world.

The use of LNG in shipping is growing with progressively more tonnage being brought online. There is a significant order book for LNG fuelled ships from across a broad range of sectors.

LNG has been transported by sea for over 50 years and has been fully de-risked for use as a fuel on both passenger ferries and cruise ships. To date there have been no major incidence involving the use of LNG.

LNG has many advantages that could be appropriate on a new build yacht. The advantages come with challenges that need to be properly addressed during the design and build phases of the project.

This paper summarises the potential benefits and the engineering challenges linked to the use of LNG as a Yacht fuel.

2. WHY USE LNG AS A FUEL

Environmental issues are becoming more and more significant in the modern world with increased legislation impacting on operations.

The use of MGO as a fuel produces significant particulate emissions and also requires mitigation of NO_x emissions within NECA zones, this entails fitting of SCRs (Selective Catalytic Reduction) or other technology to comply with IMO tier III emissions criteria. To reduce SO_x emissions Low Sulphur MGO is necessary.

SCRs use Urea to remove Nitrous Oxides from the exhaust gases. This is typically consumed at around 0.5% of fuel consumption to reduce NO_x emissions from tier II to tier III levels. SCRs have proven in service to generate significant back pressures and this makes them difficult to integrate into wet exhaust systems. The SCR must also be maintained at high temperatures to ensure its operational effectiveness and this can significantly increase fuel consumption especially at low loads.

LNG is proven to generate less NOx emissions when used in dual fuel and pure gas engines giving reductions of over 80%. There is an almost total reduction in SO_x and Particulate emission.

The lack of particulates in the LNG exhaust makes it a much cleaner fuel and consequently cleaner engine and exhaust. This significantly reduces the burden of cleaning on crew and reduces engine maintenance costs (Harteveld F 2016). The yacht will not belch out plumes of smoke as carbon deposits are not built up within the engine.

Current LNG can provide powering compliant with the emissions standards for SO_x, NO_x, Particulates and CO_2.

LNG can be significantly safer than LPG as Natural gas is lighter than air, it will naturally rise and dissipate rather than accumulate in wells within the ship.

Modern luxury yachts may have highly onerous air conditioning requirements. The regasification process requires significant heat to be exchanged. The glycol cooling loop can be integrated into the AC system significantly reducing the AC power requirements for the yacht.

LNG as a fuel is being adopted across the cruise industry and on commercial vessels trading within NECA zones. As NECA zones will extend into the Baltic and across the world the take up of LNG as a preferred fuel is expected to increase.

3. REGULATORY FRAMEWORK

There are additional risks involved with the use of liquified gaseous fuels. These include;

Explosions hazard in the case of a gas leakage, although the combustible concentration of Methane in air is limited to the range of 5-15%, the risks of an explosion must be carefully controlled.

Low storage temperature. The boiling point of Methane at atmospheric pressure is -161°C. The impact of spillages must be mitigated to ensure structures are not damaged.

Lack of familiarity of crew. Additional training is required to ensure the excellent safety record of LNG Carriers can be replicated on smaller vessels using LNG as fuel

To manage and mitigate these risks a new regulatory framework has been developed to ensure safe operation of gas fuelled ships.

3.1 IGF CODE

The International code of Safety for Ships using Gases or other Low Flash Point Fuels, was adopted June 2015 and entered into force Jan 2017.

The code provides international standards for the arrangement, installation, control and monitoring of machinery, equipment and systems to enable LNG to be used safely as a ships fuel.

LR, DNVGL, RINA and other Class societies have rule sets for the design of gas fuelled ships.

4. ENGINE TECHNOLOGY

The use of dual fuel engines provides the security of supply from existing infrastructure with the ability to run clean fuel when required. This enables an LNG yacht to maintain the operability and reach of a conventional MGO yacht with all the advantages of clean running LNG engines.

A conventional Diesel engine uses compression ignition and very high temperature combustion to provide efficient power. Modern dual fuel and pure gas engines operate on the Otto cycle. The Otto cycle has a lower combustion temperature, with the heat transfer occurring at constant volume, whereas the Diesel cycle has a higher combustion temperature with heat transfer occurring at a constant pressure. The impact of the lower combustion temperatures in the Otto Cycle is a significant reduction in the generation of NOx. Modern regulations require excess NO_x to be removed from the exhausts when inside NECA zones.

Modern dual fuel engines can operate and switch between LNG (with a small volume of LSMGO being used as a Pilot Fuel) and pure LSMGO. In doing so the operating cycle changes from the Otto Cycle to the Diesel cycle. The engine emissions generated will vary depending on the fuel used at the time.

5. LNG SPECIFIC DESIGN PROCESS

The safe use of LNG requires an increased awareness of the additional risks involved (Simpson 2016). This process starts at the concept design level and is carried through to operating the yacht in service. With the LNG yacht design process a series of Risk Based Design reviews may be required and should include all key stakeholders in the project with the express intention of improving safety issues specifically related to the individual vessel.

6. STORAGE

Low Sulphur MGO is expected to continue to be the fuel of choice for some time due to its current almost universal adoption onboard yachts, due to availability, energy density, relatively high flash point. Fuels that are liquids at normal temperatures are much simpler to handle than either solids (hard to pump) or gases (prone to dissipating). Diesel technology is highly advanced and efficient with a global supply infrastructure in place.

LNG as a fuel is more complex for several reasons. Gases are harder to handle as are very cold liquids. Additionally, LNG is not precisely defined as its composition varies throughout the world, and with time when stored (as different constituents will vaporise at different rates). The major constituent of LNG is Methane, and this may be mixed with significant proportions of more complex hydrocarbons. Due to the variable composition of natural gas it is traded in MMBTU (Millions of Britis h thermal units) this makes simple comparisons between the value of LNG and MGO more difficult.

6.1 VOLUME

The energy density of LNG by volume is less than that of MGO. 1m^3 of MGO (approximately 840 kg) is equivalent to approximately 1.4m^3 of LNG (650 kg). The volume of LNG required is not excessively larger than that of MGO, however the LNG tanks need to be highly insulated and isolated, kept away from the hull side and separated from other compartments. On a recent Houlder project the tank space holding 150m^3 of LNG and the associated re-gasification plant and piping had a total volume of 580m^3. If the whole LNG tankage space was taken up by MGO this could hold over 5 times the equivalent useful LNG.

Vessels carrying in excess of 600m3 of Fuel Oil are required by MARPOL, either to carry the fuel away from the shell or bottom plating, or comply with an Oil Outflow Calculation.

Therefore, in a yacht with a total fuel oil capacity of less than 600m^3, all the fuel capacity can be integrated into shell fuel tank and this has a limited impact on the internal arrangement. LNG volume can be considered as additional fuel without the need to move LSMGO tankage away from the shell.

As LNG is less volume efficient, its inclusion into a yacht design will increase the overall size of the yacht and this will likely limit the application of this technology. However, this increase can be limited through careful design to ensure the most effective use of spaces around the tanks.

6.2 TANK TYPES

Several tank technologies are available, these include tanks build to the shape of the ship referred to as prismatic tanks, these mostly operate at close to atmospheric pressure and cannot contain a build-up of boil off gas and this needs to be liquified and returned to the tank. Alternatively, IMO "C" type tanks are non-integral and tend to be cylindrical pressure vessels with an allowable pressure up to a maximum of 9 bar. Boil off gas is generally controlled by using the gas to fuel generators. In the case where there is no electrical load, the tank can store LNG by allowing pressure to build up. The requirement is for this method of pressure build up to enable LNG storage without vapour release for at least 15 days. For smaller applications such as those applicable to Yachts, the lack of liquefaction and simpler operation make C-type tanks the preferred option.

The tank should be maintained at or close to the boiling point of methane and should not be emptied and allowed to warm back to ambient temperatures. To do this a heel of about 10% of tank volume is left in the tank. If the tank temperature does rise, controlled cooling is required before the LNG tank may be refilled.

Prismatic (Non-cylindrical) "C" type tanks that will enable a more efficient use of space. These are under development, however they will be heavier to contain the same pressure.

6.3 TANK ARRANGEMENT

The tank will need to be arranged in a tank room and the tank room needs to be large enough to allow for inspection of all parts of the outside of the tank. This is part of the reason LNG requires a considerable volume with the vessel.

To minimise the hazardous zones all the tank connections should be contained within a tank connection space adjacent to the tank. This connection space needs to be designed for the maximum possible leakage event and design for the lowest temperatures expected in the event of an accidental release of LNG.

6.4 TANK LOCATIONS

The IGF code states that fuel tanks must be located inboard of B/5 "inboard from the ships side, perpendicular to the ships centreline". This tends to place the tank closer to amidships to maintain sufficient beam to accommodate the tank.

The tank must also be B/15 above the baseline, and at least 800mm away from the side shell

A main fuel tank may be orientated either vertically or longitudinally. A transverse orientation is not considered practicable due to the increased sloshing and small tank length.

For a new build yacht, there are a number of potential locations for LNG tankage. For a single tank solution on a yacht with a conventional arrangement with crew accommodation fitted on the lower decks forward, and stores and technical spaces below, this space would be taken over by a single longitudinal LNG tank. The impact on the arrangement would be limited to small increases draft and depth, the lower accommodation deck may need to be raised to enable the tankage to fit below it. Significant areas will remain available for technical spaces and stores outboard of the tank.

7. ATEX ZONES

The addition of gaseous fuels creates additional hazards relative to the operation of a vessel with liquid fuels. Hazardous areas are classified into ATEX zones on the basis of the frequency and duration of the occurrence of an explosive atmosphere.

ATEX zone 0 – Continuously, for long periods or frequently. Requires ATEX equipment Category 1

ATEX zone 1 – Likely to occur occasionally under normal operations. Requires ATEX equipment Category 2

ATEX zone 2 – May occur for short periods under abnormal operations. Requires ATEX equipment Category 3.

Within the ATEX zone the equipment used must be provided with an appropriate ATEX certification to ensure that it is safe to use within this zone. Typical features would include non-sparking electronic components and high temperature elements protected from gas ingress.

It is important to ensure that vent outlets are clear of high risk areas on the vessel and non-gas safe systems must be segregated and kept clear of ATEX zones

8. MACHINERY SPACES

Following the development of new technology, gas safe installations are now the preferred solution. Gas safe means a single failure in the fuel system cannot lead to the release of gas into the machinery space.

9. SYTEMS REQUIREMENTS

Several additional systems are required to be installed and these will have an impact on the yacht. These are required to ensure the yacht can operate with LNG in a safe manner.

9.1 NITROGEN GENERATION:

Nitrogen is required to purge gas fuel lines as may be necessary from time to time leaving them filled with inert gas. The system will comprise an air compressor, nitrogen generator and a receiver. The volume of nitrogen required is governed by the largest enclosed volume of the fuel lines from either the tank to the engine or the bunker station to the fuel tank.

9.2 REGASIFICATION SYSTEM

Glycol based vaporization would be required to convert the LNG to a gas, ready for combustion in the engines. This system can be linked to the yachts main HVAC system to enable cooling of accommodation. Passenger Ferries have reported significant savings in AC power requirements using this technology.

9.3 ADDITIONAL VENTILATION

In addition to the main HVAC systems and conventional machinery ventilation systems, an additional independent ventilation system will be required to ensure the safety of the gas piping and ATEX zones.

The routing of fuel pipes must be prioritised early in the design process. These fuel pipes must be kept away from the side shell. The diameter of these double walled pipes is significant, routing them can be a challenge. Most critical are the tank safety valves and fuel pipe vent pipes, these should rise in the direction of the ventilation. This is to ensure that any LNG in the vent pipe cannot create a vapour lock in the pipework leading to a pressure increase.

To ensure the safe ventilation of the high-risk spaces onboard the yacht specific "spark free" ventilation fans are required. The tank connection space will need to be maintained at an under-pressure, and isolated with air locks, that are at an intermediary pressure, to provide a double layer of protection against any gas being released into safe spaces within the yacht. To enable the under pressure of the tank connection space relative to the associated air locks these spaces shall have forced extraction and natural ventilation.

Procedures are required on board in the event of a failure in the safety critical ventilation arrangement for the gas systems. In a duel fuel vessel this may include operating in an MGO mode. As a failure in the ventilation system of the gas system will impact on safety, redundancy is required in the ventilation fans.

9.4 GAS DETECTION

Domestic natural gas includes an additive to enable detection. Natural gas without this additive is an odourless gas, LNG appears as a boiling liquid similar but colder to boiling water. To detect a gas leakage from a LNG system requires gas detectors. These are used within all areas of the vessel were gas may accumulate and are integrated into the ventilation system providing monitoring of high risk areas within the LNG system. The detectors enable timely and appropriate action to be taken in the event of a fuel leakage. For example, if the gas detection system senses a high gas concentration in the double-walled piping of an individual engine, this engine only will be changed over to MGO mode of operation automatically and the gas valve for that engine will be closed.

9.5 FIRE DETECTION

In the event of a fire detection alarm within the tank connection space or the machinery space will cause an automated shutdown in the gas system, with all running engines seamlessly changing over to MGO mode of operation.

9.6 FIRE EXTINGUISHING SYSTEMS

The application of liquids to LNG causes rapid phase transition to occur, with the LNG boiling under the liquid causing a rapid increase in Volume. This well documented hazard is avoided by the use of dry powder fire extinguishing. These must be used exclusively on LNG fires.

9.7 ENGINE SHUT DOWN AND RESTART

When an engine is operating in dual fuel mode, the gas valves will close automatically and the engine will run down in MGO mode. During the rundown period the fuel gas lines will be purged with nitrogen. Dual fuelled engines may only be started in MGO mode, with gas being introduced after a safety check cycle has been completed.

9.8 MAIN ENGINE EXHAUST SYSTEM

A ventilation system is required in the main engine exhaust to purge the system of any methane during the rundown period. Additionally, the exhausts silencers and pipework are protected from LNG explosion by relief valves or bursting discs.

10. BUNKERING

The bunkering arrangements for LNG require special consideration to ensure safety and when refuelling. It is advantageous to locate the bunkering station close to the storage tanks.

The bunkering station is defined as an ATEX Zone 1. Thus bunkering stations need to be either located on open decks or integrated within dedicated bunker stations with Shell Doors.

If the bunker stations are semi-enclosed they will be required to have forced exhaust ventilation and entry arranged via an air lock.

The bunkering system must ensure that no gas is released to the atmosphere during the bunkering process. In addition, all bunkering lines must be drained and gas freed following bunkering.

11. LNG AVAILABLITY

The availability of LNG is increasing with the increased use of LNG within commercial shipping. Currently refuelling is undertaken by bunkering barges and also by road tankers, and is becoming more available at commercial ports.

12. FUTURE DEVELOPMENTS

Recent developments in fuel cell technology mean that the potential for quiet power generation with minimal exhaust emissions from this maturing technology. This has the potential to provide complete powering without a reciprocating engine. Fuel cells run on Hydrogen and LNG is considered to be an effective source of hydrogen.

13. CONCLUSION

LNG as a marine fuel is now in use on a diverse range of vessels. and its uptake is growing. It is now in use in marine, power and rail industries. Its application to Yachting is only a matter of time.

Additional hull volume is required to accommodate a larger fuel volume. There are additional risks with the use of LNG that need additional systems to be fitted to ensure safe operation. Due to the specific requirements of LNG it will be difficult to retrofit to an existing yacht. LNG engines are an effective way of reintroducing wet exhausts due to the removal of SCRs and the clean exhaust gases will not impacting on guest areas.
Yacht owners have exacting demands on the performance of their yachts. By providing a clean fuel that reduces the HVAC burden there are very real advantages to using LNG.

14. ACKNOLEDGEMENTS

The authors would like to express their gratitude to Wartsila UK for their continued support.

15. REFERENCES

1 Fergusson Shipbuilding - Hull No 801 and 802. Currently in Build

2 Harteveld F (2016) Wärtsilä Fuel Gas Handling – RoPax Conference

3 Simpson, M. (2016) Innovation and design challenges for UK's first dual-fuel ferries. Gas Fuelled ships

16. AUTHOR'S BIOGRAPHIES

Ed Harland is a Senior Naval Architect at Houlder Limited. Ed is responsible for Naval Architecture on a broad variety of projects on yacht from 15 to 120m, this including concept design, writing technical specifications and a variety of in service support projects, including refit support, inclining experiments and stability studies. Ed's involvement with superyachts includes time working at Devonport and Sunseeker as well as consultancy projects with BMT NG and Houlder.

Jonathan Strachan is a Project Director at Houlder Limited. Jonathan is responsible for technical project management of concept and basic designs, a recent example being the basic design of UK's first LNG powered ferries, currently under construction at Ferguson Shipyard. Jonathan's experience includes design and technical consultancy for mega yacht projects of greater than 115m, new build passenger ship plan approval and survey along with shipyard design office experience.

OPTIMIZATION BASED DESIGN OF HIGH SPEED CRAFT PROPELLERS

S Gaggero, G Tani, D Villa and **M Viviani,** University of Genoa, Italy
P Ausonio and **P Travi,** DETRA Custom Propellers, Italy
G Bizzarri and **F Serra, Azimut|Benetti,** Italy

SUMMARY

The state of the art of fast propeller design codes is mainly based on classical vortical theories by Lerbs. Their crude simplifications make them often insufficiently accurate in the case of modern fast propellers geometries having highly skewed blades, non-conventional profiles, mixed type of cavitation and strict constraints about pressure pulses and radiated noise. Advanced design tools, based on (more) accurate flow solvers able to deal with the multi-objective nature of any modern design are, definitely, required. In light of this, the design of a propeller for a high-speed craft is addressed by using a multi-objective numerical optimization approach, based on a fast and reliable BEM, a parametric description of the geometry and a genetic algorithm. The new propeller is designed to improve the propulsive efficiency, reduce the cavitation extension, increase the inception speed and simultaneously maximize the ship speed. A final geometry is selected to verify the reliability of the design procedure by means of dedicated RANSE analyses, cavitation tunnel tests and full-scale measurements on a high-speed craft provided by Azimut|Benetti.

1. INTRODUCTION

The state of the art of fast propeller design codes is commonly based on classical vortical theories. These methods, all derived by the vortex line theory of Lerbs [1] with surface exact or approximated corrections, have been used and adapted in many propeller design codes. A significant step towards the design of highly-loaded or unconventional geometries was represented by the variational approach by Kerwin et al. [2], extensively adopted for the design of different types of propellers also under severe working conditions. All these approaches do take into account important design constraints, such as inception of cavitation and structural strength. However, the crude simplifications contained in these methods make them often insufficiently accurate, in case of modern fast propellers geometries having, for example, highly skewed blades, non-conventional profiles or mixed type of cavitation. Typically, the chord distribution is optimized for the design speed with some margin to prevent off-design cavitation that is only considered a posteriori, with some modifications to mitigate its negative influence in high-speed operations. On the contrary, the combination of chord and thickness should be better set through the design process to a minimum able to maximize efficiency and still prevent cavitation (sheet and midchord bubbles) in correspondence to any of the functioning conditions under investigation. In addition, the usual design tables or the curve envelopes of the minimum pressure coefficient [3] generally are not suitable when multiple design points are considered, when unconventional profile shapes are employed or when wide chord at tip are preferred. Moreover, neglecting (or approximating) the influence of the blade thickness may represent a serious drawback of a traditional lifting line/lifting surface approach in dealing with off-design conditions that may involve midchord bubble cavitation. Cavitation and unsteadiness, indeed, play the most important role in the design of high performance propellers for high-speed boats. In these vessels, propellers are often subjected to strong non-uniform inflow and the unsteady cavitation could reduce their performance, in addition to the risk of corrosion and induced noise and vibrations that certainly have to be minimized in the case of pleasure yachts. These requirements definitely require more advanced design tools, based on (more) accurate flow solvers and able to deal with the multi-objective nature of each new innovative design.

Among the potential based theories, Boundary Elements Methods (BEM) have inherent better abilities to capture the thickness effect of the blade, of the hub and, eventually, also the effect of cavitation. Panels discretizing the surfaces are placed on the real geometry rather than on simplified and approximated representations as in the case of usual low fidelity methods like lifting line/lifting surface design approaches and the inclusion of the influence of the thickness is, for instance, of particular importance when monitoring the risk of bubble midchord cavitation. The method developed at the University of Genoa [4], similarly to those developed by different research groups [5, 6] can effectively predict the steady and the unsteady flow around the propeller with sheet cavitation, including supercavitation. Nevertheless all such models, which capture (and hence require in turn) more detailed information about the propeller geometry, cannot be directly integrated into a direct design procedure, and, in fact, they are often used for validation of a given propeller design. Their level of accuracy is close to RANSE solvers [4, 6] at the design point while in very off-design conditions some discrepancies may be observed [8] but their computational efficiency is particularly high, allowing their systematic application in an inverse design process. A parametric optimization procedure represents, actually, the ideal application of BEMs for the design of such specific propellers, allowing for a more congruent and effective search of the best geometry subjected to more strict constraints and requirements. Into an optimization procedure, which can systematically change the main geometric parameters of the blades to converge on the multi-objective optimum Pareto solution, Boundary Elements Methods can be directly employed as they had been developed for, namely as analysis methods, taking advantages of all their specific peculiarities. In addition, population-based algorithms can natively handle multi-objective optimization tasks, overcoming the well-known limitations of gradient-based searching algorithms and they can exploit the computational efficiency of the BEMs for the evaluations of the thousands of solutions required for the Pareto convergence.

Such applications have already been successfully carried out in the case of conventional and unconventional propellers, including ducted, tip loaded and CLT geometries [9, 10, 11, 12].

In the present paper, an example of the application of the design by optimization to a high-speed boat propeller is given. The attention is focused on the optimization (maximization of efficiency and reduction of cavitation) of a custom propeller for a 95 feet high-speed craft manufactured by Azimut|Benetti designed with a traditional lifting surface approach. The objectives (or the constraints) on cavitation need to be revised in the light of the working conditions of such kind of high-speed propellers. Rather than requiring the minimization of the predicted cavity extension (or volume) or a sort of "advanced cavity constraints" like those successfully proposed by Vestig et al. [13], a simpler analysis of the non-cavitating pressure distributions over the blade has been preferred. The minimization of the cavity extension, indeed, was successful in the case of conventional and unconventional geometries [9, 11, 12] but, in those cases, the propellers were subjected mainly to leading edge sheet cavitation that is the kind of cavitation Boundary Elements Methods can effectively deal with. High-speed propellers, generally, are from moderately to highly loaded and the higher value of the expanded blade area, while preventing severe suction pressure at the leading edge, flattens the pressure distribution and increases the risk of michord bubble cavitation. A design by optimization based on the minimization, among the others, of the midchord bubble extension seems questionable in the light of the further approximations to be accepted within the Boundary Elements Methods to deal with this phenomenon. A simple avoidance objective (or maximization of the margin against midchord cavitation), on the contrary, seems more reliable and, being based on simpler non-cavitating calculations, even more computationally efficient. At the leading edge similar non-cavitating based criteria may successfully be derived to monitor the development of sheet cavitation too. RANSE calculations, in the light of the complexity of the flow around this kind of propellers, are, therefore, more than necessary, but their direct application as design tools in the optimization workflow, due to their comparatively lower computational efficiency is, at this moment, unfeasible. Consequently, viscous calculations in addition to measurements on a reference propeller, serve to validate the BEM, highlighting possible drawbacks and limitations to be opportunely addressed within the optimization process and, moreover, to reinforce the selection of the optimal propeller on the basis of calculations of only a reduced set of geometries from the Pareto convergence of the optimization process.

From a geometrical point of view, the design considers the widest possible variation of the main characteristics of the propeller in order to explore, within the prescribed constraints of structural strength, unconventional combinations of camber, pitch, sections mean line and chord. In addition, a different rake distribution from the one of the reference geometry is proposed. The reference rake distribution, pointing towards the suction side, was originally selected to increase the propeller efficiency at the cost of lower margin in cavitation inception [14]. The new rake distribution, pointing instead towards the pressure side, effectively works to postpone the cavity inception, at the cost however of a less significant increase in propeller efficiency. Optimization, also for this specific issue, may represent a convenient way to design a trade-off geometry, increasing its efficiency with the same margin against cavitation when the rake points to the pressure side. Complete analyses including also the original rake distribution are reported in [15]. The dedicated sea trials, discussed at the end of the paper, finally confirm the reliability and the effectiveness of this design approach.

2. THE TEST CASE

A 95 feet length ship has been used as test case. The ship shown in Figure 1, built by Azimut|Benetti group, is a pleasure craft with a maximum speed of about 26 kn. The ship is equipped with two diesel engines (MTU 16V 2000M84, MTU, 2012) delivering a continuous maximum power of 1630kW each at 2450 rpm, with two propellers in twin-screw configuration.

Table 1: Propeller functioning conditions based on self-propulsion measurements.

Speed [kn]	K_T	σ_N
21	0.247	1.16
24	0.228	1
>26 (design)	0.212	0.86

Table 1 summarizes the functioning conditions obtained from ship self-propulsion tests that will be assumed as reference for numerical and cavitation tunnel measurements while the main ship data are summarized in Table 2.

Figure 1: The Azimut 95 RPH high-speed pleasure craft.

Table 2: Main boat characteristics

Ship	Azimut 95 RPH
Design Speed	>26 kn
Length	95 feet (28.956 m)
N° Engine	2
Engine Type	MTU 16V 2000M84
N° Propellers	2
Propeller design cavitation index σ_N	0.8463

3. DESIGN TOOLS FOR OPTIMIZATION

The design by optimization relies on the application of different numerical tools. A Boundary Elements Method, developed at the University of Genoa [4, 7], is adopted to efficiently evaluate the propellers performance within the optimization process. The commercial viscous flow solver StarCCM+ [16] serves to predict hull resistance and wake, to validate the optimal blade shape candidates and to have a further insight into the complex flow features around the propellers. An optimization framework [17] and a parametric description of the geometry drive the entire design process to convergence against selected objectives and constraints.

3.1 BEM AND RANSE FOR HULL/PROPELLER PERFORMANCE PREDICTION

BEM and RANSE were used to evaluate propeller performance in both steady and unsteady (inclined shaft) functioning, with and without cavitation. The current implementation of the BEM is based on the original formulation of Morino and Kuo [18] in terms of perturbation velocity potential and Dirichlet boundary conditions. The key blade approach proposed Hsin [19] is adopted in the case of unsteady calculations (spatial non-uniform inflow or inclined shaft) while the partially nonlinear formulation of Fine and Kinnas [5] solves the sheet cavitation bubble on both suction and pressure sides. The reliability of the implementation has been verified through several benchmarks and applications, including unconventional and ducted propellers, with overall results in line with those provided by similar codes developed by other institutions and research groups. Based on these benchmarks and sensitivity analyses, calculations were carried out throughout the design by optimization process with a surface mesh consisting in 1500 hyperboloidal panels per blade and, in the case of unsteady calculations, with a time step equivalent to an angular increment of 6°. An example of the panel arrangement over the reference propeller is shown in Figure 2.

The inherent approximations introduced by the potential flow assumption prevent the application of the BEM for a reliable analysis of the propeller in off-design conditions. In addition, if for the design by optimization only the relative merits of one geometry with respect to the other (and not its absolute performance) may be sufficient for the selection of the optimal configurations, the characterization of the final performance of a propeller requires high-fidelity calculations, like those provided by RANSE solvers. In the present application, the finite volume solver StarCCM+ has been applied to overcome some of the limitations of the panel method and have a further guidance for the selection of the optimal propellers.

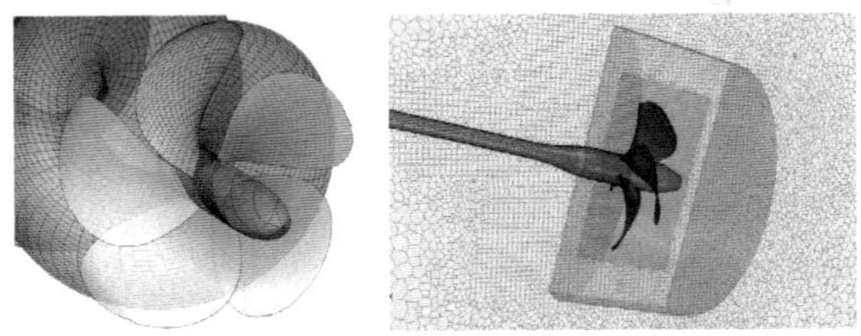

Figure 2: Surface mesh (BEM) and volume mesh (RANSE) for propeller performance analysis.

An unstructured polyhedral mesh was selected for all the calculations. Details about the boundary conditions, the most appropriate arrangement of mesh refinements and prism layers may be found in [4, 8, 20]. For open water simulations, by exploiting the axial symmetry of propeller and inflow, only a blade passage has been modelled, for a total of about 1.2 million cells per blade and an average non-dimensional wall distance of 150 to fully exploit the *realizable k-epsilon* wall functions. A snapshot of the resulting mesh and of the refining regions in inclined shaft conditions, where sliding meshes were set to allow the motion of the inner propeller region with respect to the outer domain, is shown in Figure 2. In order to perform calculations with an acceptable Courant number, an equivalent time step of 3° was assumed for the non-cavitating simulations. The inclusion of cavitation modelled with the Volume of Fluid approach within the homogeneous mixture approximation and the Schnerr-Sauer mass transfer model [21], required a further reduction of the time step, which was set to 0.5°.

As expected the agreement between measurements and calculations in open water conditions (Figure 3) is good. With respect to measurements, forces predicted by the

Boundary Elements Method are slightly underestimated, in particular for unloaded conditions. Close to the design advance coefficient (about J = 0.8) the underestimation is of 3% for thrust and 0.5% for torque. High-fidelity RANSE results are slightly better. At the design point, thrust and torque are 2.5% under- and 0.5% overestimated respectively. By assuming that the same numerical issues will affect each new geometry derived by means of small geometrical modifications of the reference propeller, a calibration of BEM results based on RANSE results was accepted for the prediction of the functioning conditions of any propeller analysed in the design process.

The prediction of cavitation in uniform inflow with BEM and RANSE at the design cavitation index is summarized in Figures 4 and 5. Advance coefficients equal to 0.7 and 0.8 and 0.9 were selected to include, by means only of uniform inflow calculations, all the possible blade sections working conditions.

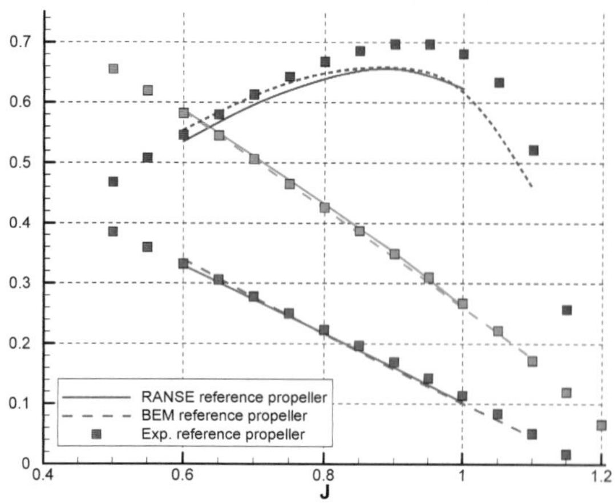

Figure 3: Comparison among measurements, BEM and RANSE calculations in open water conditions for the reference propeller.

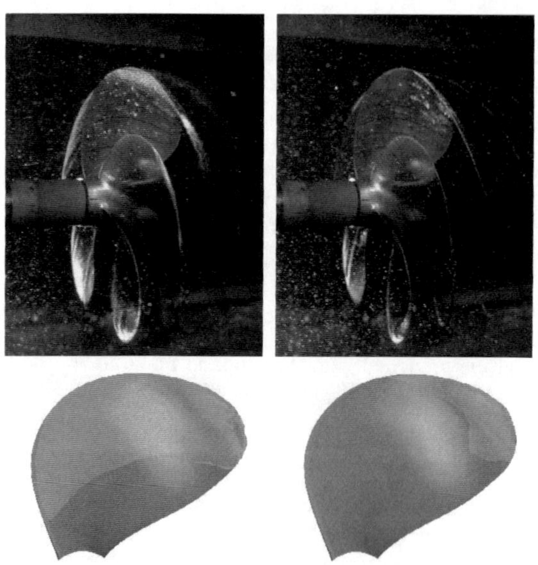

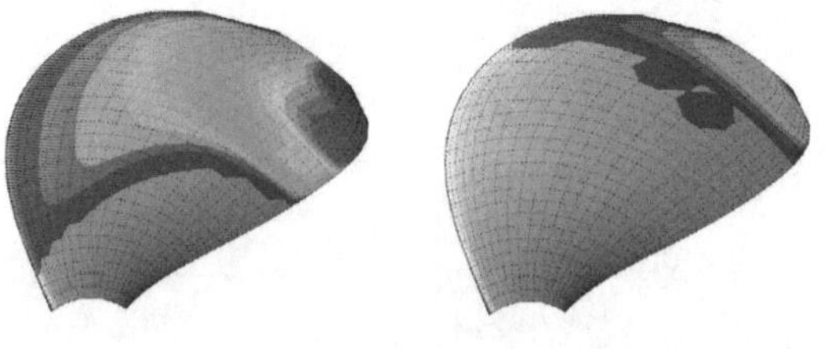

Figure 4: Cavitation tunnel observation, RANSE and BEM predictions of cavitation on the suction side, J = 0.7 (left) and J = 0.8 (right).

With the cavitation models of both the BEM and the RANSE it is not possible to discern between sheet and bubble cavitation and the presence of bubbles at midchord is approximated with a continuous, very thin, sheet vapour bubble over almost all the blade suction side (loaded condition, J = 0.7). At J = 0.8, the agreement between RANSE and BEM is satisfactory and the predicted spots of vapour in the tip region close to the trailing edge are well in agreement with the observations. In lightly loaded condition (J = 0.9) also BEM foresees midchord bubbles that however underestimate the radial extension observed during measurements and reasonably evidenced by RANSE. Furthermore, pressure side cavitation at leading edge, observed at J = 0.9, is similarly predicted by the two numerical approaches.

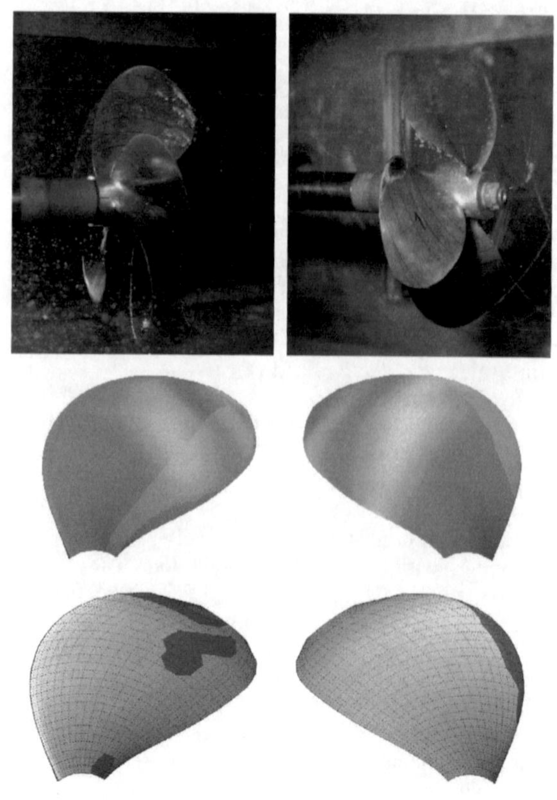

Figure 5: Cavitation tunnel observation, RANSE and BEM predictions of cavitation, J = 0.9 on the suction side (left) and J = 0.9 on the pressure side (right).

RANSE calculations were used also to predict the relevant flow characteristics around the fully appended hull, in order to have an estimation of the resistance curve in still water conditions and of the nominal wake on the propeller plane to verify its unsteady functioning. An example of calculation is shown in Figure 6. Free surface was predicted using, again, the Volume of Fluid approach without mass transfer on a hexa-dominant mesh (2 Million cells for half the hull) properly refined with anisotropic spacing in correspondence of the free surface. The resulting nominal wake is that of Figure 7. By analysing the vertical component of the flow at 90 and 270° positions, the equivalent shaft inclination (combination of geometrical shaft inclination, hull dynamic attitude and local flow features on the stern of the ship) may be estimated in 11°. In axial direction, the only disturbance is represented by the narrow decelerated zone at 12 o'clock position, reasonably generated by the shaft line and its appendages. These features of the wake field may influence the development of unsteady cavitation but the effect of the inclined shaft is definitely the most important one: the average and the radially varying distributions of flow at 90° and 270° may represent significant functioning conditions for approximated, quasi-steady analyses while the shaft inclination (both experimentally at the cavitation tunnel and numerically with the BEM and the RANSE) is a reasonable simplification of the functioning of the propeller for additional verifications of the numerical tools adopted in the optimization.

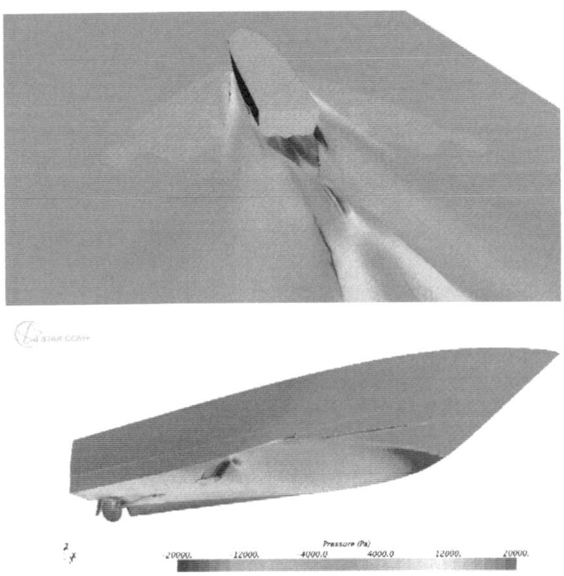

Figure 6: Free surface elevation and pressure distribution over the hull by RANSE calculations.

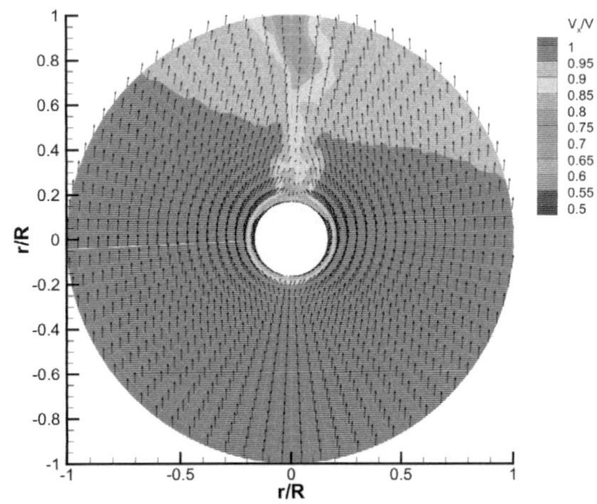

Figure 7: Predicted nominal wake on the propeller plane by RANSE calculations.

3.2 PARAMETRIC DESCRIPTION OF PROPELLER GEOMETRY

The shape of a propeller blade is traditionally described by curves that represent the most significant distributions (pitch, chord, maximum camber, thickness, among the others) from root to tip. Such representation is inherently parametric and may be directly used for the automatic generation of the geometries to be analysed within the optimization process. Fairness and smoothness are the principal requirements of the description. In the present work, a classical modification of the geometrical distributions of the propeller, described with B-Spline curves, as deeply discussed in [9, 11, 12] was preferred to these aims. The limited number of control points of these curves turn into the free parameters of the optimization, reducing the number of decision variables but still allowing for complex geometry variations and facilitating the fairness of the shapes. Similarly, also local variations of the sectional profiles of the blade may be included in the design process thanks to an identical description of the chordwise distributions of sectional thickness and camber.

In the present case, a set of 48 free parameters was selected to alter the reference propeller geometry. The ranges of variation were selected to balance the need of exploring unconventional shapes and combinations with the need to avoid unrealistic performance improvements due to unfeasible shapes that results from the lack of physics in the medium-fidelity code represented by the BEM. This choice consists in:

- 7 parameters to describe the chord distribution,
- 8 parameters to describe the sectional pitch distribution,

- 9 parameters to describe the maximum camber distribution,
- 12 parameters to describe the B-surface that represents the non-dimensional camber surface of the blade,
- 12 parameters to describe the B-surface that represents the non-dimensional thickness surface of the blade.

3.3 CONSTRAINTS AND OBJECTIVES

Optimization is inherently a multi-objective process and, being the design of a propeller always a trade-off between competing objectives and constraints, the Pareto convergence among which select the best compromise can be better addressed if all requirements and constraints can be turned into parameters and outputs that the codes adopted for the design may accurately evaluate.

With the exception of the geometrical constraints represented by the limits forced for the free variables and hard-coded in the design space, the only other two constraints accepted in the proposed design process regard the blade strength and the maximum speed achievable by the ship.

Any change of chord and pitch results in a change of the load and, in turn, of the functioning point of the propeller. The thickness distribution, therefore, is not free to change independently. A minimum sectional thickness, further corrected to comply with classification rules [22], is derived for any design by approximating the blade to a beam, subjected to the actual thrust and torque, rigidly fixed at the root.

The second constraint concerns the operative conditions of the propeller. In general any new design has to assure a ship speed as close as possible to the reference geometry and forcing a certain delivered thrust for the geometries defined by the optimization fixes this operative point. The fulfilment of an equality constraint, however, is a serious bottleneck for the design process, even if certain margins were accepted to avoid an excessive number of unfeasible geometries and speed up the convergence process. To overcome this limitation, in the present work, a different approach has been adopted, converting the thrust constraint into a couple of objectives: instead than requiring a given thrust, the new propellers have been designed in order to maximize the boat speed at the engine Maximum Continuous Rate. The type of the boat, obviously, allows for this change, being the increase of the maximum cruise speed always an added value for the customer. The optimization workflow requires, consequently, specific procedures. For each geometry it is necessary to evaluate the correlation between rate of revolution, required power and maximum achievable speed (possible by assuming unchanged the propulsive coefficient measured for the reference propeller) that turn into an objective to be maximized while minimization is required for the "distance" of the actual functioning point from the MCR (as for Figure 8). At the resulting maximum and endurance ship speeds, the risk of cavitation is evaluated and assumed as a further set of design objectives..

Computational efficiency is the key point of the design by optimization and simplified calculations and design criteria are of fundamental importance when thousands of geometries have to be tested in different functioning conditions. The design proposed in this work is based on rather crude assumption (whose reliability however has been accurately verified in the case of the reference propeller) that anyhow closely resemble well-established design procedures but included in a completely automatic design environment.

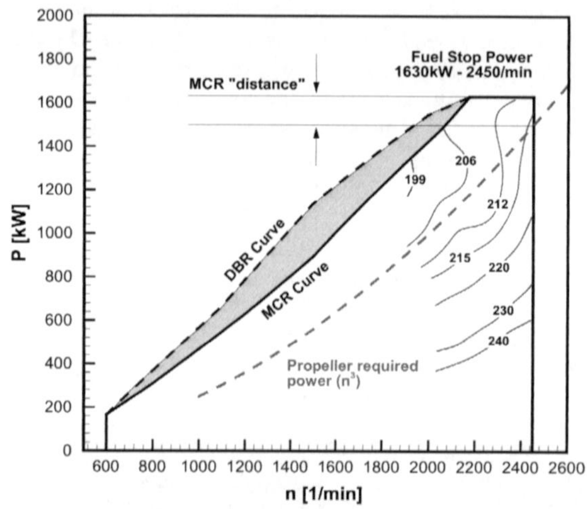

Figure 8: MTU 16V 2000M84 Engine layout and required propeller power. Maximum speed and minimum distance from the MCR are the additional objectives of the design.

Quasi-steady analyses with radially varying inflow in correspondence of the few critical positions inside the hull wake preliminary computed with RANSE (90° and 270° blade position for loaded and unloaded conditions respectively plus the average inflow) approximated the unsteady behaviour of the propeller in inclined shaft, following the successful application of this approximation of Bertetta et al. [9].

Furthermore, non-cavitating analyses were preferred, even if in previous optimization activities the use of the cavitating model with the BEM provided successful results [9, 10, 11, 12].

The risk and the possible margins against cavitation can be monitored, indeed, by using the simplest criterion on pressure ($-C_{PN}<\sigma_N$) for any points on the blade with a single non-cavitating calculation while the type of cavitation (sheet or bubble) can be marked by the position, along the section chord, in correspondence to which the dynamic pressure falls below the vapour pressure. Especially for what regards bubble cavitation, as already mentioned, a simple avoidance criterion based on the minimum of the pressure distributions may result more computationally robust.

In addition, the expected extension of the cavitating region, with non-cavitating analyses, can be controlled by the combination of two criteria: the portion of the pressure diagram below the vapour pressure (herewith addressed as "inception area", being the region were cavitation for sure starts and occurs) and the strength of the suction peak. Larger zones of "inception area" with higher suction peaks (as per Figure 9) are a symptom of more extended cavitation that, consequently, can be handled in the design process by requiring their minimization. A combination of the two, a kind of "inception volume", may be considered as a measure of the cavitation strength.

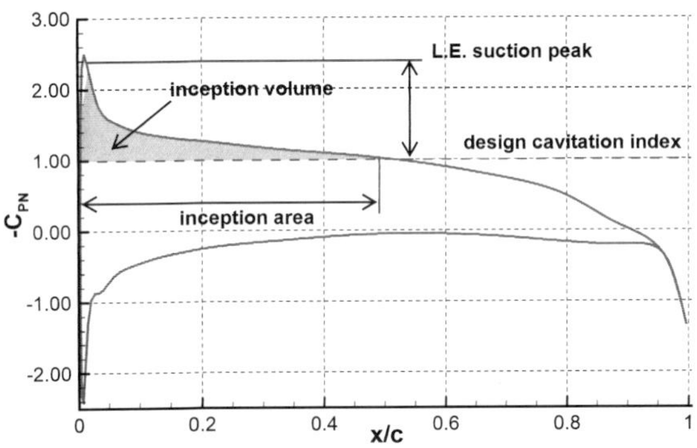

Figure 9: Monitors of the cavitating behaviour of each propeller blade radial section: "inception area", "inception volume" and pressure suction peaks.

The comparison of the inception areas of the reference propeller (indifferently BEM or RANSE) of Figure 10 with the fully developed cavity extensions of Figures 4 and 5 demonstrate the robustness and the reliability of the simplification, especially if compared to cavitation models that poorly approximate bubble phenomena. Inception areas by both BEM and RANSE reasonably foresee, at the different functioning conditions under investigation, the presence of vapour observed at the cavitation tunnel and predicted with the cavitation model turned on, confirming a positive correlation of the phenomena (inception area criterion, fully developed bubble and, more importantly, observations) to be exploited in the design by optimization process. The large area where the pressure coefficient is below the design cavitation index persists, at midchord near the tip, also when the local angle of attack is reduced. This is a symptom of "flat" pressure diagrams at the design point that underline the risk of bubble cavitation (observed and approximated by the cavitation models) which may represent a possible margin of improvement of an already well designed propeller. Both the chordwise and the radial extensions of the inception areas underestimate the presence of cavitation, whose "true" extension (as shown by the calculations with cavitation enabled) is, of course, significantly affected by the development of the bubble itself. The analysis proves, anyhow, that the simpler non-cavitating pressure distributions can fairly serve as an indicator of the different types of cavitation and of their risk of occurrence in the design process.

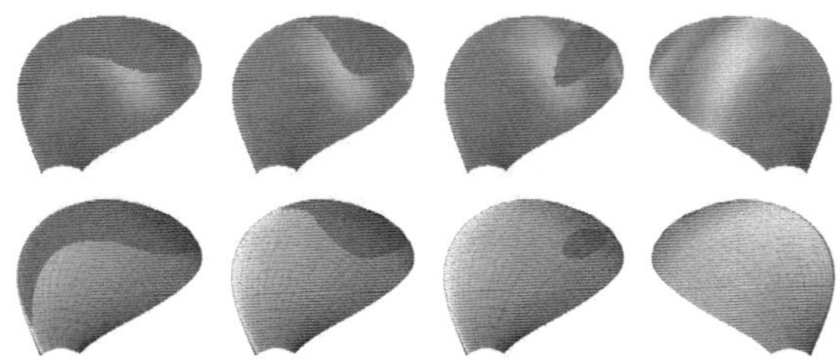

Figure 10: Inception areas (RANSE on the top, BEM on the bottom) for the reference propeller at J = 0.7, 0.8, 0.9 and 0.9 on the pressure side.

The design of the new propeller based on these simplified assumptions was, thus, carried out by requiring:

- Maximization of the boat speed at the maximum delivered power,
- Minimization of the distance of the working point from the engine maximum continuous rate point,
- Maximization of the propeller efficiency at the endurance speed (70% MCR)
- Minimization of the cavitation "inception area" on the suction side - equivalent flow at 90° blade position at maximum delivered power/speed
- Minimization of the cavitation "inception area" on the pressure side - equivalent flow at 270° blade position at maximum delivered power/speed
- Minimization of the "inception volume" on the suction side - equivalent flow at 90° blade position at maximum delivered power/speed
- Minimization of the "inception volume" on the pressure side - equivalent flow at 270° blade position at maximum delivered power/speed
- Minimization of the leading edge suction pressure peak coefficient for any blade sections at the tip ($r/R > 0.7$) at maximum delivered power/speed
- Minimization of the leading edge suction pressure peak coefficient for any blade sections at inner radii ($r/R < 0.7$) at maximum delivered power/speed
- Minimization of the midchord suction pressure peak coefficient for any blade sections at maximum delivered power/speed
- Minimization of the cavitation "inception area" on the suction side - equivalent mean inflow at endurance speed
- Minimization of the "inception volume" on the suction side - equivalent mean inflow at endurance speed

In addition, also the inception areas on the pressure side at maximum speed at the loaded condition (90°), on the suction side at maximum speed at the unloaded condition

(270°) and on the pressure side at the endurance speed for the average flow were monitored and all the geometries that showed any of these issues were discarded during the post-processing of the results.

3.4 OPTIMIZATION WORKFLOW FOR PROPELLER DESIGN

By combining the above mentioned parametric description of the geometry, the appropriate flow solvers and by selecting objectives and constraints, the design by optimization can be seen as an automatic "trial and error" procedure, that in present case is driven to convergence by using a multi-objective genetic type optimization algorithm within the ModeFRONTIER software environment [17]. The design of the optimal geometry is indirect, rising from the analysis of hundreds (thousands) of different possible configurations among which the better candidates against the selected constraints and objectives may be identified.

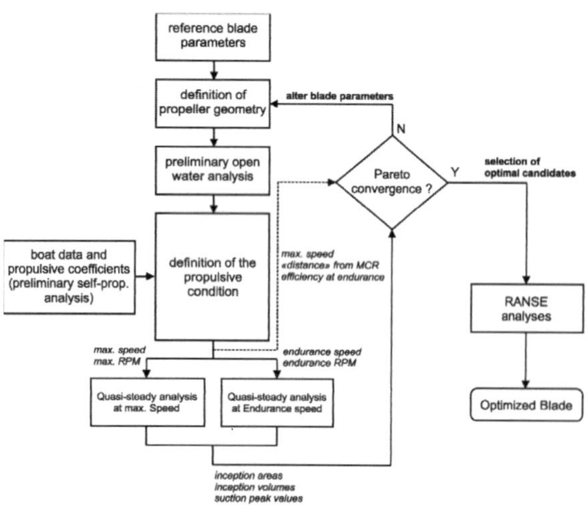

Figure 11: Workflow of the design process based on optimization.

Due to the large amount of free parameters and objectives it was necessary to select an initial populations sufficiently wide (1500 members) to reasonably cover the design space and let it evolve for 35 generations to achieve a satisfactory Pareto convergence. BEM, for these reasons, results as the only possible medium-fidelity computational approach to be employed for the evaluation of the about 50000 geometries iteratively defined by the optimization algorithm in a time (about 7 days on a common workstation) compatible with the usual design process. RANSE calculations, as shown in the flow chart that summarizes the design process of Figure 11, were adopted only a posteriori for a reanalysis of a set of Pareto configurations (including inclined shaft calculations with cavitation) among which identify, on the basis of further detailed high-fidelity analyses, the optimal geometry.

4. THE OPTIMIZED PROPELLER

The preliminary analysis of the reference propeller confirms the complexity of the design of highly-loaded high speed propellers. Margins for the improvement of the cavitating performance are rather limited with almost "shock-free" pressure diagrams at the design point (Figure 16). A change in the rake distribution to unload the tip [12, 14] could increase cavitation margins at tip.

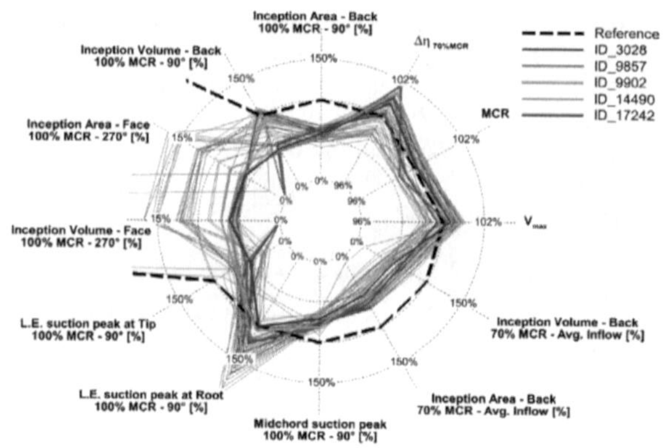

Figure 12: Sub-selection of Pareto candidates.

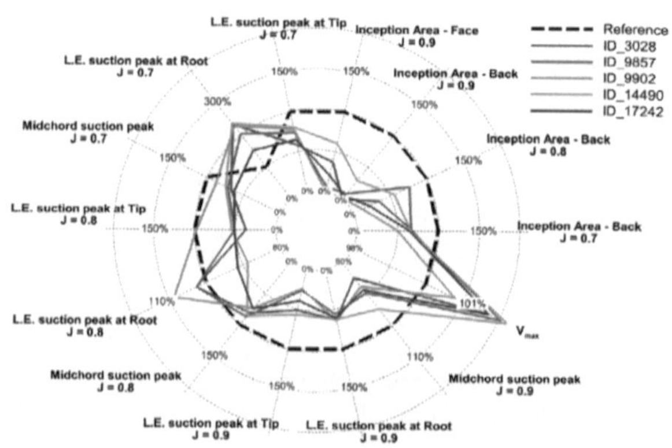

Figure 13: RANSE performance for the selected geometries.

On the other hand, the required maximization of the boat speed turns into a differently distributed load with a decrease of efficiency and an increased risk of sheet cavitation at intermediate radial positions that the optimization process has to mitigate.

A summary of the most significant geometries identified by the design process based on BEM is shown in Figure 12. Among these, five (ID_3028, ID_9857, ID_9902, ID_14490 and ID_17242) have been selected for further RANSE analyses, whose results are summarized in Figure 13. Only a selection of the most significant design objectives has been included in the representation of results.

The effectiveness of the design process is remarkable. The unloading of the blade due to rake distribution determines a consistent reduction, among the Pareto designs and in particular for the selected geometries, of the inception area at the maximum loading in correspondence to the 90° blade position. Also the suction peak at tip is significantly lowered as a symptom of retarded inception, as well as the inception area on the pressure side with few geometries even succeed in completely avoiding face cavitation. These improvements, obviously, have some costs. As a result of the redistribution of load induced by the different rake distribution and the simultaneous maximization of the boat speed and propeller efficiency (that on the basis of BEM calculations increased up to 1.5% and 2% respectively), the suction peak at on blade sections close to root is higher with a consequent increased risk of sheet cavitation. These results by the Boundary Elements Method are completely confirmed by the RANSE. The reduction of the inception area at any working conditions evidenced by the BEM in the optimization process is exactly computed when viscous calculation in "equivalent" uniform inflow (J=0.7 instead of 90° blade position, J=0.8 instead of average inflow and J=0.9 instead of 270° blade position) are considered. For all the five geometries under analysis, the improvements in terms of extension of the region with a pressure below the vapour one are remarkable and he inception points at different loading have been successfully postponed.

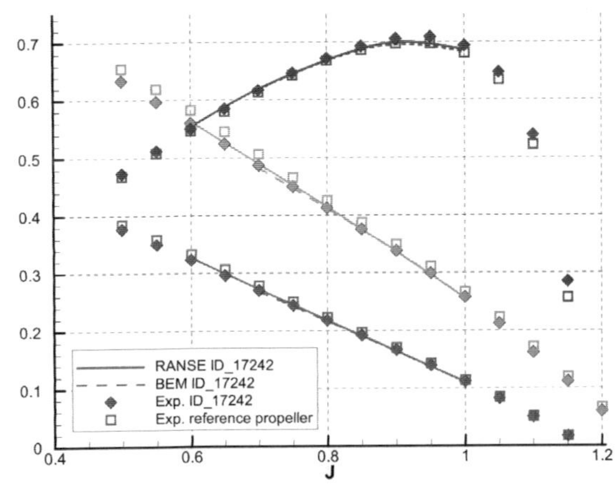

Figure 14: Measurements, BEM and RANSE calculations in open water conditions for the ID_17242 propeller. Comparison with measured thrust and torque of the reference propeller.

Only the suction peak at root in the loaded condition (J = 0.7) is significantly higher than the reference propeller, as anticipated by the Boundary Elements Method analyses; this has been voluntarily accepted since focus of optimization is the elimination of midchord bubbles. ID_17242, based on BEM and RANSE preliminary calculations, provides a significant reduction of the cavity extension with the overall better performance in terms of postponing of the cavity inception (at loaded and design conditions), with the exception of the unloaded condition where ID_14490 and ID_3028 perform slightly better. For these reasons, it was selected for detailed numerical analyses and dedicated cavitation tunnel observations, as well as full-scale surveys.The comparison between towing tank tests and numerical calculations of Figure 14 confirms, in addition to the analyses carried out for the reference propeller, the reliability of both the BEM and the RANSE approaches. For the whole range of advance coefficients under investigation, the accuracy of both calculations is even higher than that observed in the case of the reference geometry, in particular for what regards the Boundary Elements Method, which in general suffers rake distributions excessively pointing towards the suction side, as that of the reference propeller. Close to the design point, around J = 0.8, the error with respect to measurements is lower than 0.5%: this, in turn, should imply the reliability of the maximum craft speed prediction, within the assumed hypotheses of invariance of the propulsive coefficients as a consequence of a different propeller geometry. Measurements show a slight increase of the propeller efficiency that even in the simpler case of open water conditions is in line with outcomes of the numerical calculations carried out for the optimization.

RANSE predictions of the pressure distributions at the three advance coefficients representative of loaded, design and unloaded functioning of the propeller operating in inclined are shown in Figures 15, 16 and 17.

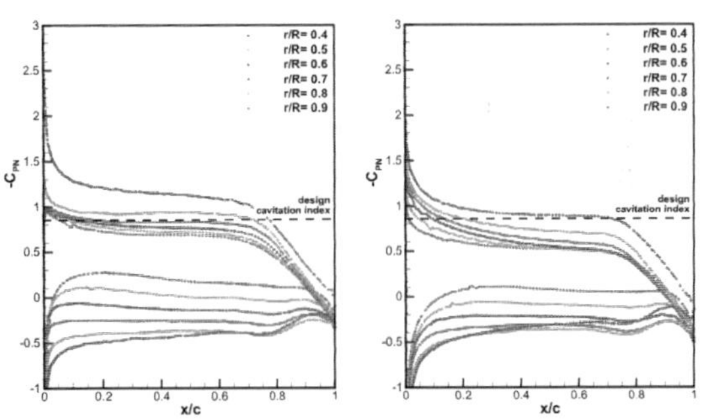

Figure 15: Pressure distributions at different radial sections. RANSE calculations for the equivalent J = 0.7 functioning condition. Reference propeller (left) and ID_17242 (right).

Even if the inception point is slightly postponed, the optimized propeller still shows a consistent risk of leading edge sheet cavitation at J = 0.7, as expected by the unloading of the tip and the radial redistribution of load. As expected, the ID 17242 however delivers its best performance close to the design point and in unloaded conditions. At J = 0.8 the risk of cavitation is limited to the tip of the blade with a certain margin against bubble cavitation anywhere else on the blade. In correspondence to the unloaded condition, the suction side is completely free of cavitation: on the pressure side only few sections are subjected pressure below the vapour limit, the inception is postponed and a non-negligible margin protects from midchord bubble cavitation.

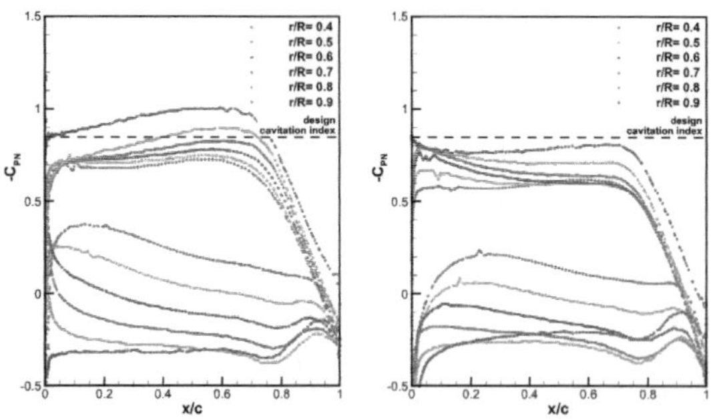

Figure 16: Pressure distributions at different radial sections. RANSE calculations for the equivalent J = 0.8 functioning condition. Reference propeller (left) and ID_17242 (right).

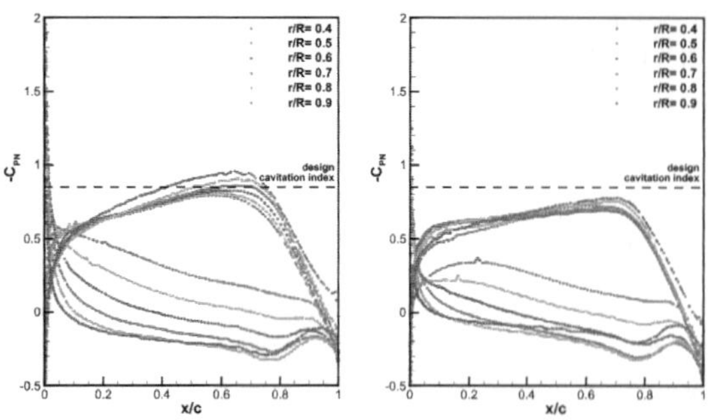

Figure 17: Pressure distributions at different radial sections. RANSE calculations for the equivalent J = 0.9 functioning condition. Reference propeller (left) and ID_17242 (right).

These results are furthermore sustained by the experimental observations in inclined shaft proposed in Figure 18. As expected, bubble cavitation is significantly reduced. Only at 90° the optimized propeller shows some midchord bubbles that are, however, far from the extension observed on the reference blade and in accordance with the outcomes of the design process. The absence of sheet cavitation on the pressure side in the case of the optimized propeller proves the reliability of the numerical calculations. Cavitation inception points are summarized and compared in Figures 19 and 20 for sheet/vortex and bubble phenomena respectively. The optimized propeller shows substantial improvements for any of the phenomena observed with the exception of bubble cavitation at root in correspondence to the 270° position that is a further proof, by comparing the Pareto diagrams of Figures 12 and 13, of the reliability of the criteria adopted during the optimization. The reduction of cavitation positively influences also the delivered thrust. With respect to the reference geometry (-5.7%), the optimized propeller has an almost negligible thrust breakdown (-0.4%) in cavitating conditions that may balance the less effective role of the pressure side rake in achieving higher efficiency. Numerical calculations (both BEM and RANSE in inclined shaft) mainly overestimate the reduction of thrust in cavitating conditions. This is particularly true in the case of the reference propeller (-11% with RANSE) for which, however, the peculiarity of the cavitating phenomena dominated by midchord bubbles, may represent a valid explanation of the poor performance of cavitating models not completely suited for these conditions.

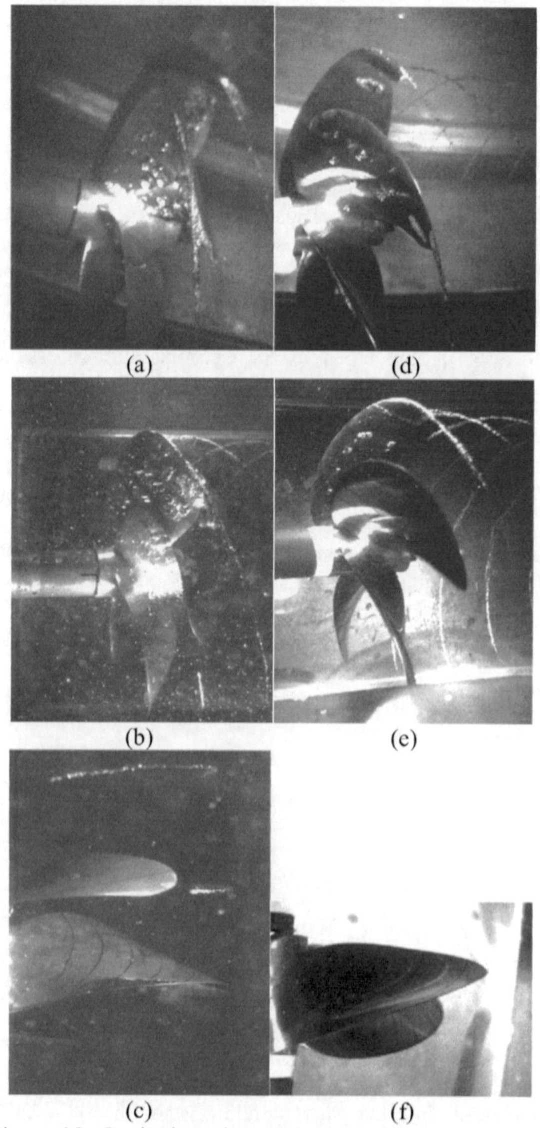

Figure 18: Cavitation observations. Reference propeller: (a) suction side at 90° from starboard, (b) view from above and (c) pressure side at 270°. ID_17242 propeller: (d) suction side at 90° from starboard, (e) view from above and (f) pressure side at 270°.

Trends (-2.1% of thrust reduction for the reference propeller computed with RANSE), which are qualitatively in agreement with measurements, still confirm the effectiveness of the design by optimization. The reliability of the design and the consistency of the improvements foreseen by the optimization is finally verified by the dedicated sea trials summarized in Table 3.

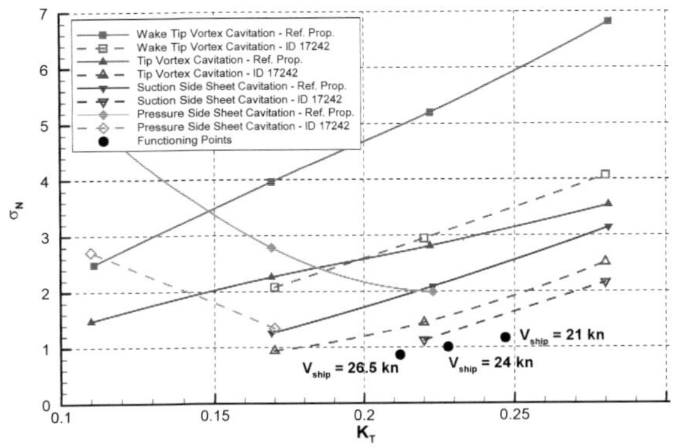

Figure 19: Comparison of the experimental inception point of vortex and sheet cavitation.

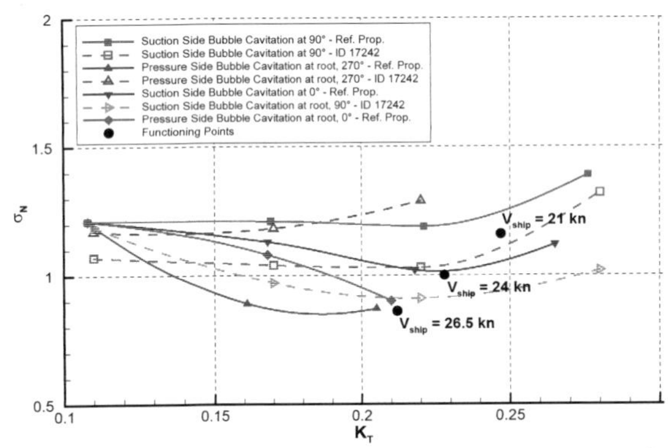

Figure 20: Comparison of the experimental inception point of bubble phenomena.

Table 3: Sea trials results at 100% of the delivered power.

		Max. Speed [kn]	Propeller Rate of Revolution [% MCR]	
			Left Shaft	Right Shaft
Reference Propeller	Run 1	26	98.9%	98.3%
	Run 2	25.3	98.7%	98.7%
	Average	25.7	98.8%	98.5%
ID_17242	Run 1	26.2	98.7%	99.9%
	Run 2	27.1	98.9%	100.0%
	Average	26.7	98.8%	100.0%

The increase of 1 kn is higher than that predicted during the optimization using BEM (+0.25kn) and in the post-processing of the selected geometries using RANSE (+0.5kn). These differences, further to the uncertainties of measurements, could also be explained by the approximate inclusion of the actual thrust breakdown in the design process by using quasi-steady BEM analyses that only the final RANSE calculations can reasonably predict. This can be considered as one of the key point still open for an even more robust automatic and computationally efficient design process.

5. CONCLUSIONS

In the present paper, a design by optimization process, based on genetic algorithm and BEM calculations, has been proposed and validated by using dedicated model scale measurements and sea trials for a high-speed propeller. The devised procedure relies on some simplifications (inception area concept and quasi-steady calculations, need of day affordable designs) whose effectiveness has been analysed considering viscous computations and, preliminarily for the reference propeller, also experimental measurements. Suitable formulations of the optimization objectives and constraints have been, consequently adopted. An explicit blade strength estimation has been included in the process, together with an engine – propeller matching algorithm to derive the functioning conditions of each geometry defined by the optimization and to change usual thwarting constraints circa the delivered thrust into design objectives.

After the analysis of the Pareto candidates and additional high-fidelity RANSE calculations on a rationally selected subset of geometries, an optimal propeller was identified. Model scale experiments largely confirmed the reliability of the design tools for what regards both the prediction of propeller characteristics and the estimation of cavitation as well: efficiency is higher and cavitation sensibly reduced. From this point of view, cavitation tunnel experiments furthermore clarified the bubbly nature of most of the suction side cavitation phenomena predicted in the form of sheet cavitation by both the numerical methods due to the inherent limitations of the cavitating models. The easier analysis of non-cavitating pressure distributions turns out to be, consequently, a valid simplification to be exploited in the design process. Sea trials surveys, finally, confirmed the overall reliability of the tools employed in the optimization and, even if accounting for the small differences between the two crafts equipped with the propellers, they still represent an important indication of the effectiveness of the proposed design procedure. The need of more accurate predictions of thrust breakdown due to cavitation results, on the other hand, as a mandatory aspect for an even more robust automatic and computationally efficient design process.

6. ACKNOWLEDGEMENTS

These activities have been carried out in the context of a cooperative project among Azimut|Benetti, DETRA Custom Propellers and the University of Genoa. And have been partially funded by the MiSE – Ministero dello Sviluppo Economico, within the ABC (Advanced Boat Concept) Project – (Progetto n. C01/0889/00/X19 - Decreto di concessione del MiSE 02308 del 23 dicembre 2013).

7. REFERENCES

1. LERBS, H.W., 'Moderately loaded Propellers with a Finite Number of Blades and an Arbitrary Distribution of Circulation', *SNAME Transactions, 60*, 1952.

2. KERWIN, J.E., CONEY, W.B. and HSIN, C.Y., 'Optimum circulation distribution for single and multi-component propulsors', *Proceedings of Twenty-First American Towing Tank Conference*, 1986.

3. BROCKETT, T.E., 'Minimum Pressure Envelopes for Modified NACA-66 Sections with NACA a=0.8 Camber and Buships Type I and Type II Sections', *DTMB Report 1780*, 1966.

4. GAGGERO, S., VILLA, D. and BRIZZOLARA, S., 'RANS and panel methods for unsteady flow propeller analysis', *Journal of Hydrodynamics, 22*, 2010.

5. FINE, N.E. and KINNAS, S.A., 'A Boundary-Element Method for the Analysis of the Flow around 3-D Cavitating Hydrofoils', *Journal of Ship Research, 37(3)*, 1993.

6. VAZ, G., 'Modelling of Sheet Cavitation on Hydrofoils and Marine Propellers using Boundary Element Methods', *PhD Thesis, Technical University of Lisbon – IST*, 2005.

7. BRIZZOLARA, S., VILLA, D. and GAGGERO, S., 'A systematic comparison between RANS and Panel Methods for Propeller Analysis', *Proceeding of 8th International Conference on Hydrodynamics ICHD2008*, 2008.

8. GAGGERO S., VILLA D. and VIVIANI M., 'An investigation on the discrepancies between RANSE and BEM approaches for the prediction of marine propeller unsteady performances in strongly non-homogeneous wakes', *ASME 2014 33rd International Conference on Ocean, Offshore and Artic Engineering*, 2014.

9. BERTETTA, D., BRIZZOLARA, S., GAGGERO, S., SAVIO, S. and VIVIANI, M., 'CPP propeller cavitation and noise optimization at different pitches with panel code and validation by cavitation tunnel measurements', *Ocean Engineering, 53*, 2012.

10. GAGGERO, S., RIZZO, C.M., TANI, G. and VIVIANI, M., 'EFD and CFD design and analysis of a propeller in decelerating duct', *International Journal of Rotating Machinery*, 2012.

11. GAGGERO, S., GONZALEZ-ADALID, J. and PEREZ SOBRINO, M., 'Design of Contracted and tip loaded propellers by using boundary element methods and optimization algorithms', *Applied Ocean Research, 55*, 2016a.

12. GAGGERO, S., GONZALEZ-ADALID, J. and PEREZ SOBRINO, M., 'Design and Analysis of a new generation of CLT propellers', *Applied Ocean Research, 59*, 2016b.

13. VESTIG, F., GUSTAFSSON, R., and BENSOW, R. E., 'Development and application of optimisation algorithms for propeller design', *Ship Technology Research, 63(1)*, 2016.

14. DANG, J., 'Improving Cavitation Performance with New Blade Sections for Marine Propellers', *International Shipbuilding progress, 51(4)*, 2004.

15. GAGGERO, S., TANI, G., VILLA, D., VIVIANI, M., AUSONIO, P., TRAVI, P., BIZZARRI, G. and SERRA, F., 'Efficient and multi-objective cavitating propeller optimization: an application to a high-speed craft', *Applied Ocean Research, 64*, 2017.

16. CD-ADAPCO, 'StarCCM+ 9.06.011 Users Guide', 2014.

17. ESTECO S.R.L., 'ModeFRONTIER 2014 User's Manual', 2014.

18. MORINO, L. and KUO, C.C., 'Subsonic Potential Aerodynamic for complex configuration: a general theory', *AIAA Journal, 12*, 1974.

19. HSIN, C.Y., 'Development and analysis of Panel Methods for Propellers in Unsteady Flows', *PhD thesis, Massachusetts Institute of Technology*, 1990.

20. GAGGERO, S. and VILLA, D., 'Steady cavitating propeller performance by using OpenFOAM, StarCCM+ and a boundary element method', *Proc. IMechE Part M: Journal of Engineering for the Maritime Environment*, 2016.

21. SCHNERR, G. and SAUER, J., 'Physical and numerical modeling of unsteady cavitation dynamics', *Proceedings of the 4th International Conference on Multiphase Flow*, 2001.

22. RINA - Registro Italiano Navale, 'Rules for the Classification of Yachts Designed for Commercial Use', *Part C, Chapter 1, Section 7*, 2013.

8. AUTHORS BIOGRAPHY

Stefano Gaggero holds the current position of Researcher at the University of Genoa. He is responsible for propulsor flow application and associated development with BEM and RANSE codes.

Giorgio Tani holds the current position of Researcher at the University of Genoa. His main interests are focused on the experimental fluid dynamics, cavitation tunnel measurements, and acoustic measurements.

Diego Villa holds the current position of Researcher at the University of Genoa. His main interests are focused on Computational Fluid Dynamics for both propellers and ships and the associated code development.

Michele Viviani holds the current position of Associate Professor at the University of Genoa. He is responsible for the experimental activity at the cavitation tunnel of the University of Genoa. His main interests are focused on experimental fluid dynamics, noise, cavitation tunnel measurements, propeller design and optimization.

Pierluigi Ausonio is the responsible of propellers design at DETRA Custom Propeller. He has a thirty-year experience on the design of custom propeller for a wide range of ship types and sizes.

Piero Travi is the head of DETRA Custom Propeller. He has a thirty-year experience on the design of custom propeller for a wide range of ship types and sizes.

Giovanni Bizzarri holds the current position of Director of the R&D Department of Azimut|Benetti and of Director of the megayacht technical office.

Francesco Serra is a Naval Architect and Marine Engineer in the R&D Department of the Azimut|Benetti group. He is involved in Computational Fluid Dynamics calculations for both hulls and propellers.

RELIEVING YACHT DESIGNERS' CHALLENGE BY INTEGRATING AZIPOD® PROPULSION AND ONBOARD DC GRID DISTRIBUTION: INCREASED SAFETY AND COMFORT, CONSIDERABLE FOOTPRINT SAVINGS (OVER 40%) AND REDUCED ENVIRONMENTAL IMPACT

K Kokkila and **J Erkintalo,** ABB Marine & Ports, FI

SUMMARY

Increased requirements for safety, comfort and environment support the use of podded propulsion and advanced power distribution with batteries and other energy storage devices. However, the challenge for the yacht designer is how to fit all new requirements, especially energy storage, into the vessel design considering the limited machinery spaces in yachts.

ABB's latest technologies relieve the yacht designers' challenge in several ways. Azipod® D propulsion integrated with Onboard DC Grid distribution not only saves a considerable amount of energy, but reduces the rating and size of onboard power equipment overall – freeing up space to incorporate energy storage. This paper highlights the advantages of Azipod D propulsion and Onboard DC Grid combination such as safety and comfort of the people onboard, energy consumption, and Polar Class notations. In addition, it demonstrates in detail how weight saving and over 40% machinery footprint saving is achieved compared to conventional AC electrical power distribution with shaftline propulsion.

NOMENCLATURE

AC	Alternating current
AVR	Automatic voltage regulator
DC	Direct current
DP	Dynamic position
ES	Energy storage
GHG	Greenhouse gases
GT	Gross tonnage
CAU	Cooling air unit
NOx	Nitrogen oxide
V	Volt
A	Ampere
SCR	Selective catalytic reduction
SRU	Slipring unit
THD	total harmonic distortion
JB1	Junction box 1

1. INTRODUCTION

Two alternative electrical propulsion and power distribution layouts were defined for non-specific but a representative mega yacht vessel to compare several attributes between the two alternatives. These alternatives are defined based on ABB's market leader experience gained throughout the 25 years in mega yachts' electrical propulsion business.

The base case alternative, named AC-SHAFT, is a traditional AC power plant combined with two shaftlines with electrical onboard propulsion motors. Steering is made with two conventional rudders.

The alternative layout, named DC-POD, is based on ABB's DC power distribution solution, Onboard DC Grid, and has two Azipod DO1250P units for propulsion and steering instead of shaftlines and rudders, see Fig 1.

Both of the alternative layouts have, in addition to propulsion load, a hotel load of 1,000 kVA and battery capacity of 820 kWh.

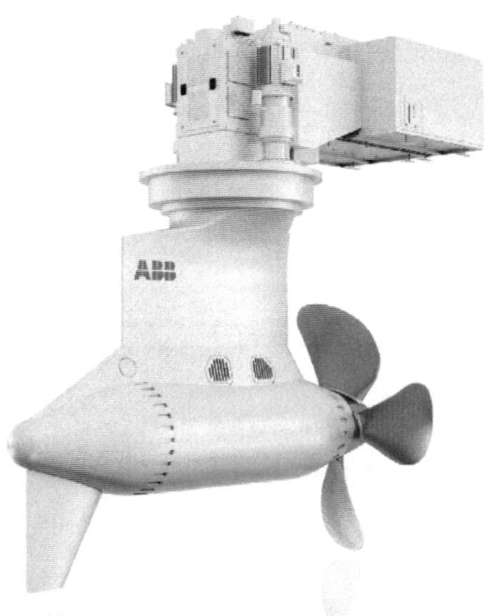

Fig 1. Azipod DO propulsor is a fully azimuthing propulsor with pulling propeller providing low noise & vibrations

2. AZIPOD PROPULSION

The selected propulsion power for the base case AC-SHAFT is 2 x 3500 kW. Typically, pulling pod propulsion decreases the power need from 5 % to 10 % compared to shaftline-rudder in the case of a twin propeller vessel with limited draft such as a ferry, RoPax, yacht or cruise ship. This reduction is gained from better propulsion efficiency and smaller resistance if the hull lines are optimised for Azipod propulsion.

The first twin shaftline vs. twin Azipod propulsion study was a full-scale comparison of sister vessels in the late 90s. For the first six vessels of the 70,000 GT Fantasy-class cruise ship series, electric propulsion and two conventional shaftline propellers were installed. The last two vessels of the series were installed with twin Azipod propulsion without conducting any major optimisation for the existing shaftline based hull lines. In that case, the Azipod equipped vessels achieved 9 % improvement in propulsion efficiency [1].

The more recent comparison was made by ship designer Deltamarin Ltd for ABB Marine & Ports in Dec 2016. The comparison was based on Deltamarin's recent 20,000 GT RoPax design with 2 x 5 MW shaftline propulsion including contemporary flap rudders. The base case had been tested in a model basin. They redrew the hull lines for twin Azipod DO propulsion and made a comparison of both alternatives in CFD, see Fig 2 and Fig 3. This resulted in an 11 % decrease in effective power in favour of Azipod propulsion optimised hull lines. The power prediction at 20 kts design speed was for original shaftline version 2 x 5.0 MW and for Azipod version 2 x 4.3 MW.

2.1 HULL RESISTANCE

Decreased hull resistance with Azipod propulsion comes from several factors. First typical pod advantage is in skeg thickness and removal of stern thruster and their openings. Second, hull lines for shaftline versions sometimes have partial tunnel shapes to direct water to the propellers. These can be removed from Azipod versions and the straightened-out shapes lead to a further decrease in resistance.

Fig 2. Optimised hull lines for conventional twin shaftline-rudder propulsion and steering.

Thirdly, the slope of the vertical lines can be optimised as well with Azipod propulsion, since there are no restrictions on maximum available shaft length and bearing distance. This have particularly large impact on vessels with a low block coefficient, e.g. passenger vessels. A gentle slope angle has itself a direct influence on the viscous resistance by equalizing the dynamic pressure changes around the hull, but it also has several derivative influences, such as a thinner boundary layer, which mostly affects propeller design and thus leads to better wakefield and propulsive efficiency as well.

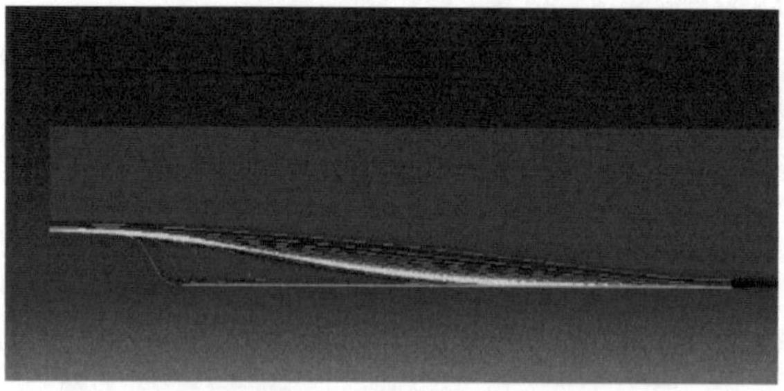

Fig 3. Optimised hull lines for twin Azipod propulsion.

2.2 PROPULSIVE EFFICIENCY

Azipod propulsion has added benefits when moving from effective to delivered power. Pulling propellers receive a steady undisturbed wake field, which gives propeller designers a greater scope to optimize propellers for higher efficiency and/or silent operation, see Fig 4.

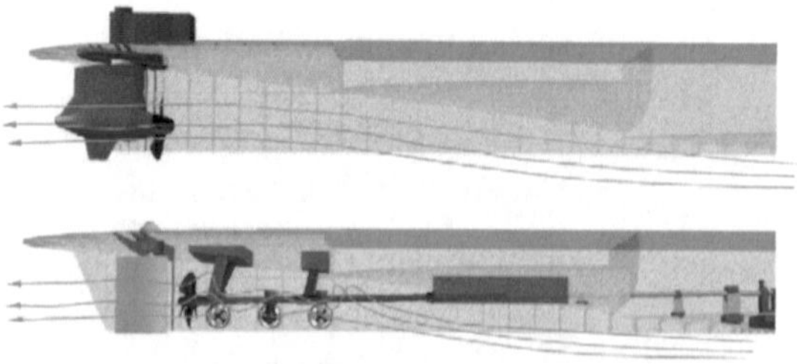

Fig 4. Azipod propellers receive undisturbed wake in optimum angle.

In addition, Azipod propulsors can be inclined forward to be in line with the local flow, typically 2-4°. Conventional shaftline propellers on the other hand are typically inclined backwards 1-1.5°. This means that shaftline propeller blades are 5-7° off the direction of inflow, which has a negative impact on the propulsive efficiency, vibration, cavitation and noise performance.

Even though it is quite unusual to have exact propulsion power comparisons between shaftline and Azipod propulsions with real ship models developed and optimised for the same project, we are confident to assume some propulsion power reduction for the DC-POD alternative in order to have a realistic comparison between the alternatives. Considering the aforementioned examples and ABB Marine & Ports' know-how from hundreds of sales projects and delivered ships, we select the conservative estimation of 2 x 3,300 kW propulsion power for DC-POD alternative, meaning only about 5 % power reduction compared to base case AC-SHAFT.

2.3 OPERATIONAL SAFETY BENEFITS [2]

2.3 (a) 150 % more side thrust

With Azipod propulsion, the full propeller thrust can be directed freely in any direction, whereas in fixed shaftline-rudder arrangements thrust decreases rapidly as helm angle increases. Generally, a conventional rudder can produce only about 40 % side thrust compared to maximum ahead bollard pull thrust. The figure for flap rudders is up to 60 %. With a 360° freely turning Azipod propulsor, however, full thrust can be precisely applied in any direction, giving 150 % more side thrust than a conventional rudder. Furthermore, Azipod propulsion makes it possible to navigate astern and sideways simultaneously, which is difficult to achieve with a rudder since negative propeller speeds reduce the effectiveness of a rudder considerably.

2.3 (b) 38 % smaller turning circle

An Azipod-equipped vessel is more likely to avoid collision than a vessel with a conventional shaftline-rudder arrangement. This is because conventional rudders typically require tunnel thrusters at the stern to assist weak manoeuvring. However, tunnel thrusters do not work effectively at higher ship speeds, whereas the superior steering capability of Azipod units is effective throughout the ship's speed range. Furthermore, Azipod units eliminate the need for stern tunnel thrusters, thus providing greater flexibility and simplicity in ship design and general arrangement.

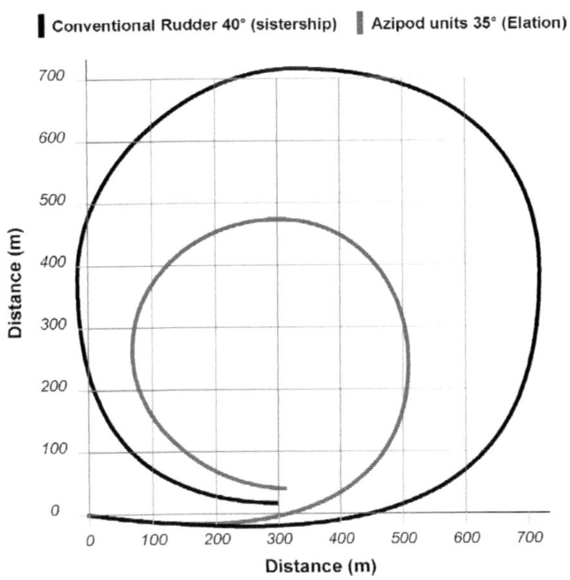

Fig 5. Full-speed steering tests from MS Elation show the superiority of Azipod steering compared to her Fantasy class rudder equipped sister vessel

More effective and safer turning capability of Azipod propulsion has been verified, for example, by full-scale and full-speed turning circle tests between sister-ships MS Fantasy with conventional propulsion and MS Elation with Azipod propulsion which recorded 38% reduction in tactical diameter [1], see Fig 5. Similar results have been obtained also from model experiments with a wider set of ships, see Fig 6.

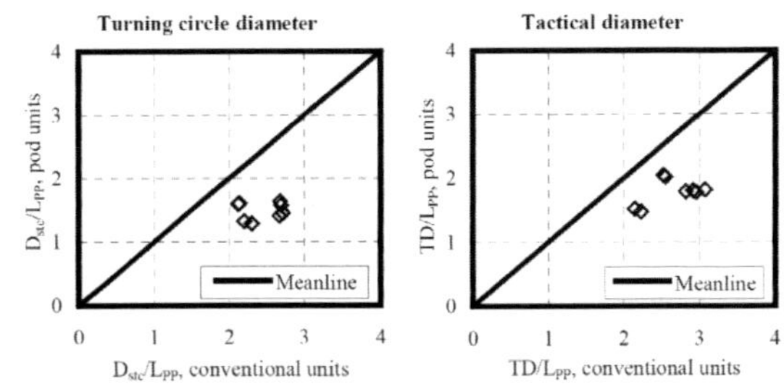

Fig 6. Turning circle data of ship models equipped with podded or conventional propulsion units [3]

2.3 (c) 50 % shorter crash-stop distance with full heading control

With traditional rudder steering, an emergency crash-stop is accomplished by reversing the propeller rpm from positive to negative. This is time-consuming as the ship power generating machinery must go from full power to zero power and then ramp up again to full power in the opposite direction. In practice, the vessel operating with a rudder will also lose her heading control during the crash-stop as the rudder will not work efficiently unless the propeller is producing thrust for it – and at negative propeller rpm there is little thrust available for the rudder at all.

This means that ship heading and direction during the crash-stop are effectively at the mercy of the elements – determined by prevailing environmental conditions, i.e. current, wind and waves, which especially becomes a factor in heavy seas. In Azipod vessels, the crash-stop can be accomplished in the "pod-way" by steering the Azipod units outwards 180 degrees and keeping positive propeller rpm during the whole crash-stop. This shortens the crash-stop distance considerably – typically by at least 50 %, see Fig 7. Moreover, during the crash-stop, Azipod units can generate enormous side force in any desired direction irrespective of ship speed. This gives the captain full control to decide the heading and direction of the vessel during the whole crash-stop even in heavy weather conditions.

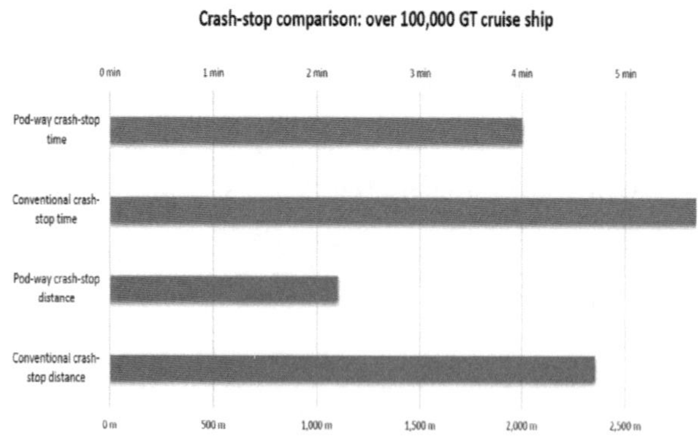

Fig 7. Full-scale comparison between "pod-way" and conventional crash-stops from full speed

2.4 ICE CLASSES [2]

The growing popularity of the Arctic and Antarctic as destinations for tourists has brought a spike in orders for passenger ships capable of operating in icy waters. While most of these inquiries and projects are related to expedition cruise ships, nowadays some of the mega yacht inquiries contain requirements for polar ice class as well.

Overall, Azipod propulsion is the current global market leader in independently navigating ice-going vessels. Currently there are more than 60 very high-ice class vessels with Azipod propulsion in operation or ordered.

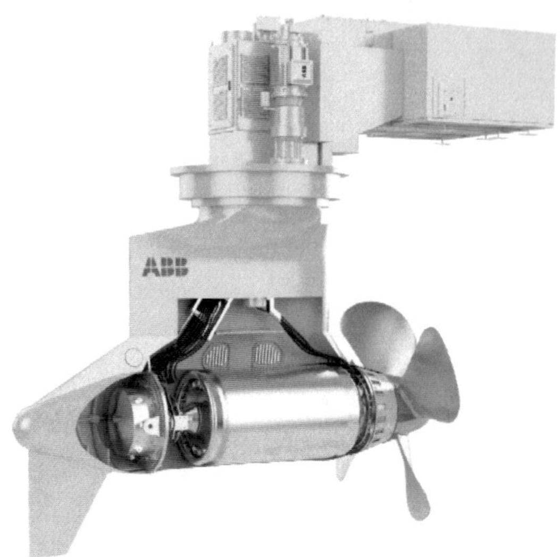

Fig 8. Azipod propulsion motor is integrated onto propeller shaft making the drive train gearless and robust.

Within the Azipod unit, the electric motor is installed directly on the propeller shaft making the drivetrain extremely simple and robust against any ice loads hitting the propeller, see Fig 8. In contrast to mechanical Z- or L-drive azimuthing thrusters, there are no mechanical gears so the Azipod shaftline can withstand both bending and high torque peaks under heavy ice loading.

In comparison to shaftline propulsion, a further Azipod safety feature derives from its steering system, which is set to yield and absorb extreme impact loads on propeller side blades. This sort of extra protection cannot be achieved for the blades of the conventional fixed shaftline propeller. This phenomenon has been verified by full-scale measurements in extreme Arctic ice conditions.

Azipod DO product series, suitable for mega yacht propulsion, covers power range from 2 MW to 7.5 MW per propeller and is available up to PC 5 ice class. The first Polar Class vessel with Azipod DO propulsion will be *Scenic Eclipse* with PC 6 ice class notation granted by Bureau Veritas (see Fig 9). Scheduled to launch in August 2018, *Scenic Eclipse* is a discovery yacht able to navigate in both Arctic and Antarctic waters.

Fig 9. The Scenic Eclipse discovery yacht has PC 6 ice class and Azipod DO main propulsion.

Another Azipod-equipped Polar Class passenger ship series will be three *Endeavor* class mega-yachts for Crystal Cruises with PC 6 notation. They are designed for operation in the Arctic, Antarctic and also in the Tropics. These 20,000 GT new buildings will be the world's largest expedition yachts with ice class and the first passenger ships to be classified according to the new DNV GL rules.

2.5 NOISE & VIBRATIONS [2]

In the Azipod unit, there are no noise-generating gears and the pod motor and its shaft are located completely outside the ship hull. Azipod unit's pulling propeller receives a steady undisturbed wake field, which gives propeller designers greater scope to optimize propellers for silent operation compared to a conventional pushing propeller with rudder, see Fig 4.

Disturbing vibrations caused by maneuvering in ports with high rudder angles are also avoided as the Azipod propeller and its housing rotate as a single unit, meaning there is never a high angle of attack between them. Noise and vibration caused by stern tunnel thrusters are avoided as these are no longer necessary when adopting Azipod propulsion.

The aforementioned aspects are one reason why Azipod propulsion features are so commonly used as a solution on large cruise ships since Comfort Class requirements are easier to fulfill than would be the case with conventional shaftline-rudder propulsion arrangements.

3. ONBOARD DC GRID

3.1 DC GRID BASIC PRINCIPLES

Onboard DC Grid used in DC-POD alternative is a modular DC based system platform tailored to the needs of the next generation of marine power train systems. Variable frequency converters are widely used in modern marine power systems, often representing upwards of 80% of the installed load. Recent developments in DC protection methods have enabled a re-arrangement of the way these loads are connected to the main system. This approach distributes power on DC in order to streamline system topology and simplify integration of energy sources like variable speed generators and energy storage.

DC Grid meets the needs of the next generation of vessels more efficiently by distributing power on DC instead of AC voltage.

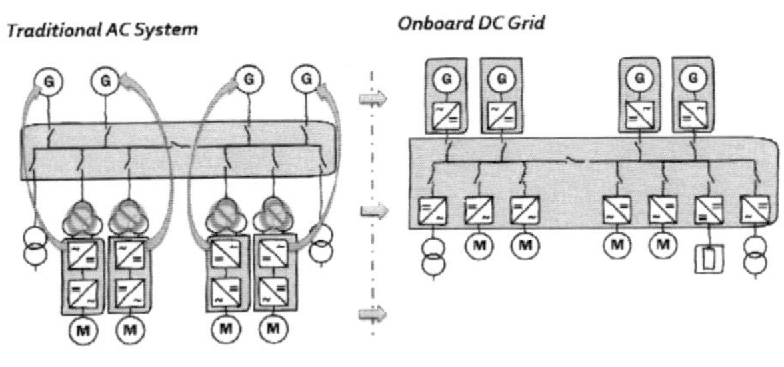

Fig 10. From traditional AC to Onboard DC Grid system configuration.

In a DC Grid system, the voltage from AC generators is rectified (AC to DC converter) immediately at the terminals of the generator and from there the power is transmitted to consumers on a DC bus. Inverters (DC to AC converter) are used to convert the power to variable frequency AC voltage in the case of variable speed motor drives or fixed frequency AC voltage when the distribution network is created. The created AC distribution network is a low THD network and conventional controlling of AC system THD level is not a problem anymore. Energy storage is connected to the system, either via a DC/DC converter or directly to the DC link. The reduced number of components in the system results in a reduced footprint of up to 30% and over in some cases as compared with an AC system.

The Onboard DC Grid platform is comprised mainly of products with long track-records from marine AC power systems. The novelty lies within the way these known products are put together in a system, and the way in which this new system has been purpose built for the marine propulsion application.

Complementing the power system is an integrated and harmonized control and communication infrastructure that facilitates a transparent and near-instant transmission of information between system components. This ensures a holistic approach to the task of coaching the best performance, be it for safety or efficiency, out of the Onboard DC Grid.

3.2 ONBOARD DC GRID EFFICIENCY & FUEL SAVINGS

Unlike AC based distribution systems where connected generators need to match system voltage and frequency, the DC Grid system only requires the generators to match system voltage. This means that the generator and engine speed can be dynamically optimized to the system load situation. When the engine load decreases, the engine speed is also reduced.

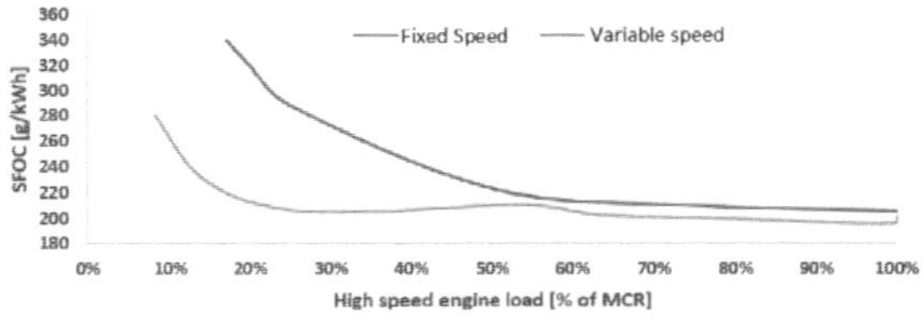

Fig 11. Variable vs fixed high speed engine SFOC curves.

The most immediate benefit of this change is reduced fuel consumption, as visualized in the graph above. From the graph can see that the fuel consumption is the same with fixed and variable speed on higher loads, but on partial loads the reduced fuel consumption is up by max. 35 %. This supports vessels that have a variable speed operational profile especially well, for example vessels that operate in DP-mode. There are also additional benefits to variable speed operation: reduced GHG emissions due to lower fuel consumption and reduced particle emissions due to cleaner combustion. Increased temperature of exhaust gases at lower loads means that SCRs can be fully operational at all load levels, reducing both NOx emissions and urea consumption. Furthermore, the lower engine rpm's reduced maintenance costs due to reduced wear and tear on the engine.

3.3 ONBOARD DC GRID SAFETY & COMFORT

DC is inherently simpler than AC. When building up a system platform this means that it is easier to predict fault scenarios and devise effective protection against them:

- Closed bus operation in DP 2 without additional equipment because common mode faults like governor and AVR failures are handled more effectively.

- With high speed engines the generators can be online in close to 10 seconds because they don't need to wait for synchronization.

- Engines that are virtually impossible to overload even when operated at lower speeds. This is because each generator has built-in overload protection that limits output power. The end result is that the engine does not stall and remains online.

- Clearing of major short-circuit currents in a "soft" way so that the system recovers quickly and predictably. The system is therefore not plagued by ugly transients as is often the case in AC systems when large fault currents are interrupted by protective devices.

Due to the variable speed engines the vessel is potential to achieve 5 dB reduction noise level equivalent to a 30-35 % decrease in perceived noise. This is achievable due to the lower rpm's in partial loads when the engine is not running at full speed during low load situation. The actual noise reduction level depends of the chosen engine characteristic.

3.4 ONBOARD DC GRID WITH ENERGY STORAGE

Energy Storage and associated technologies have received a dramatic increase in attention in recent years, not least in the maritime industry. Whilst this can be attributed to a number of different factors, what is certain is that it has the potential to improve safety, efficiency and performance of the vessels.

However; the availability of ES only gets you part of the way to improved safety, efficiency and performance. First; the energy must be made available to consumers. Since most Energy Storage media are DC based, the integration into a DC distribution system becomes simpler than doing the same into an AC based distribution system.

The DC solutions require less equipment in general (as shown in in the diagram above) and the converter (if used) also becomes significantly more compact than its AC counterpart. For basic functions where no selectivity or starting scenarios are considered, the AC converter solution is almost twice the length of the DC equivalent. If you also consider selectivity, overvoltage and starting scenarios then this ratio becomes closer to four.

The option of connecting the ES directly to the DC link offers a slight reduction in length compared to the converter option. However, this is at the expense of controllability of the current in and out of the ES and system voltage level. From a control perspective this option means that ES power flow is determined by the sum of the actions of all other sources and consumers in the system. This means that this method of connection is only suited to a limited number of applications, typically systems of low complexity where batteries represent a dominant power source.

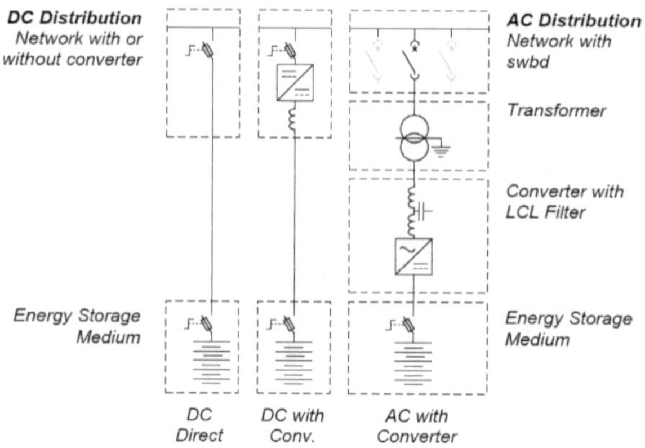

Fig 12. Energy storages integration to the grid.

The Onboard DC Grid is an open platform which easily allows reconfiguration in the number and types of power sources. This enables future usage of alternative renewable energy sources such as solar, wind and fuel cells and those can be as easily integrated to the DC grid system. With these the DC/DC converter connection method is used for the integration and transformers are not needed.

4. FOOTPRINT COMPARISON

DO-POD and AC-SHAFT alternatives were compared taking into account the footprint and weight of the main power plant and propulsion equipment, see Appendix 1. The comparison shows the individual equipment footprint and weight saving for the main power plant and propulsion items based on a realistic mega yacht case with marine classified equipment. It does not take main engine dimensions or weight into account. The result is that the DC-POD alternative reduces onboard footprint needed for main power plant and propulsion equipment 46 % compared to AC-SHAFT alternative.

The weight comparison, considering only fully comparable power plant equipment, resulted in a 27 % saving with DC-POD alternative. A more extensive weight comparison between Azipod propulsion and shaftline propulsion with rudder steering and differences in hull steel weight due to different structures is left out as this would require detailed hull structure planning. However, based on the feedback from ship designers, our general expectation is that for electric propulsion vessels Azipod propulsion is considerably lighter compared to shaftline propulsion with the same thrust, especially when taking into account savings in hull steel weight and conventional shaftline components. Even without rudders and shaftlines of AC-SHAFT, the sole onboard propulsion motor is already heavier compared to complete Azipod unit, as seen on Appendix 1.

The DC-POD alternative, see Appendix 2 for single line diagram, includes four same frame size variable speed generators with integrated rectifiers. Generators are connected to two separate drive line-ups via breakers. These drive line-ups are connected together with bustie breakers and the systems are selective with each other. Each drive line-up includes a propulsion inverter for Azipod units and distribution network island converters. Line-up 1 includes DC/DC converter for the energy storage control and line-up 2 includes inverter for the bow thruster motor.

The AC-SHAFT alternative, see Appendix 3, includes four same frame size generators with one smaller frame size hotel generator. Generators are connected to two separate main switchboards with AC circuit breakers. The main switchboard includes feeders for the multidrive. The multidrive includes inverters for the shaftline propulsion, stern thruster and bow thruster. The main switchboard also includes feeder cubicles for the distribution networks and energy storage system. The energy storage system includes a transformer and converter unit.

4.1 ONBOARD DC GRID SPACE SAVING

Onboard DC Grid system in DC-POD alternative, as presented in Appendix 2, provides multiple ways to save space in the vessel. This is due to reduced number of components in the grid system and the efficient way of integrating other systems into the grid. Some of these space saving methods are following:

- Main switchboard is not needed, this is replaced with drive line-up that includes inverters and needed breakers for disconnecting. Rectifier units are located in the main generator housing, so additional propulsion drive rectifier supply units are not needed as is needed in AC system propulsion drive.

- Variable speed generators create highly flexible power generator system. This enables improved flexibility with fewer of generators. Meaning that, when AC system requires additional generator to create smaller load step for example for hotel load, in DC Grid the same load adjustment can be done with one generator by controlling the speed.

- As mentioned in the previous chapter, energy storages integration to the grid does not require transformer or switchboard cubicle and some energy storage configuration can be connected straight to the DC link without DC/DC converter between.

- Multiple frequency converter feeders can be added to drive line-up to feed bigger and smaller consumers for example pumps and HVAC drives. This reduces the size of the distribution network components.

- Traditional AC feeder cubicles circuit breakers are replaced with a fused disconnectors that requires less space.

4.2 AZIPOD PROPULSION SPACE SAVINGS

With twin Azipod propulsion, the most notable space-saving comes from the missing shaftlines and propulsion motors onboard the vessel. The stern tunnel thruster and its electrical motor and drive can be left out as well.

Onboard propulsion motors are almost always two decks high, thus taking up a lot of space. In our case vessel, the saved machinery room footprint by removing propulsion motors alone is already 30 m^2. Volume-wise, this would be about 100 m^3 when taking into account the height of the motors. This does not include the savings from the area utilized for shaftlines and re-arrangement of machinery spaces.

The negative aspect of Azipod propulsion is that more space is needed in the steering gear room above the pods compared to conventional rudders. This is due to SRU, which transfers electrical power to azimuthing podded motor, and larger steering gear needed for Azipod, which is a combined propulsion and steering unit. The footprint needed for the Azipod DO1250P unit selected for DC-POD alternative is 8 m^2. For conventional rotary vane type steering gear, we estimate that footprint would only be about 2.5 m^2.

In such yacht designs, where there would be more space in the steering gear room, it is possible to select one step smaller Azipod DO1100P unit equipped with CAU located in the steering gear room and connected to the Azipod unit. CAU boosts the maximum power of the podded propulsion motor, thus making it possible to use a smaller Azipod frame size. However, CAU would further increase the onboard footprint in the steering gear room of the DC-POD alternative to 14 m^2.

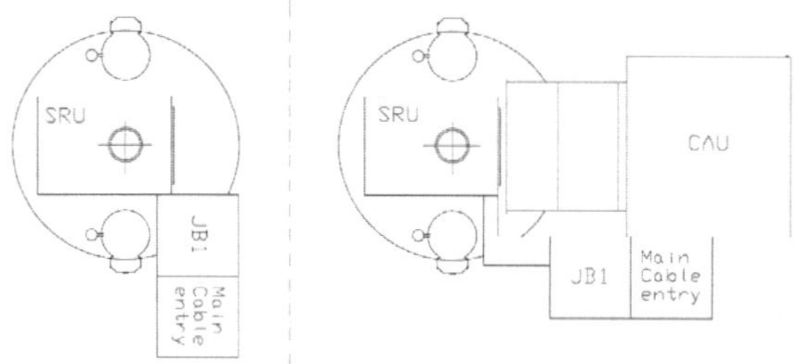

Fig 13. Azipod DO design without CAU (on the left) reduces onboard footprint considerably compared to the CAU-boosted design (right).

The auxiliary equipment which can be located quite freely in the steering gear room, e.g. steering drives and control cabinets, require an additional 2 m^2 footprint with Azipod propulsion. This we assume is similar with conventional steering gear, which also needs steering control cabinets and often have a hydraulic power unit instead of an electrical steering drive. Thus, it is left out of the comparison.

5. CONCLUSION

Based on the detailed comparison as listed in Appendix 1, the DC-POD alternative reduces the onboard footprint needed for the main power plant and propulsion equipment 46 % compared to the AC-SHAFT alternative. Most of the footprint saving with the DC-POD alternative is gained by being able to remove the following equipment:

- Onboard propulsion motors
- Main switchboards
- Shaftlines
- Harbour generator
- Aft tunnel thruster motor, drive and transformer

The weight saving, considering only fully comparable power plant equipment, is 27 % with the DC-POD alternative. We expect that weight saving with DC-POD would be further increased if the detailed hull structure would be designed for both alternatives including, for example, rudder, sterntube, shaftline and its supports, brackets and bearings for AC-SHAFT alternative.

Considering onboard noise and cavitation, DC-POD have advantage as the propeller design is easier to optimise for pulling Azipod propeller working in an undisturbed wakefield compared to a pushing shaftline propeller. In addition, high attack angles when manoeuvring with rudder and use of stern tunnel thruster are avoided with Azipod propulsion. Further, thanks to variable speed engines in Onboard DC Grid, the vessel is able to achieve 5 dB noise level reduction in partial loads, which is equivalent to 30-35 % decrease in perceived noise.

DC-POD alternative is also better in terms of environmental impact. Fuel oil consumption is typically lower thanks to (i) higher efficiency of Azipod propulsion and (ii) Onboard DC Grid distribution, which have higher efficiency especially on partial loads. In addition, particle emissions are decreased due to cleaner combustion with variable speed engines.

6. REFERENCES

1. KURIMO, R., 'Sea Trial Experience of the First Passenger Cruiser with Podded Propulsion', Practical Design of Ships and Mobile Units, page 743, 1998.

2. KOKKILA, K., 'Polar Expedition Ships: Ultimate passenger safety and comfort with Azipod® propulsion', ABB Group, Jan 2017, https://library.e.abb.com/public/06e3e581db7f4e868190bbebfc22b559/ABB_whitepaper_Polar_Cruise_Kokkila_2017.pdf, visited 4 Apr 2017.

3. TOXOPEUS, S. and LOEFF, G., 'Manoeuvring Aspects of Fast Ships with Pods', 3rd International EuroConference on High-Performance Marine Vehicles HIPER'02, Bergen, 14-17 Sep 2002, page 398, http://www.marin.nl/upload_mm/6/e/2/1806690851_1999999096_Manoeuvring_aspects_of_fast_ships_with_pods.pdf, visited 5 Apr 2017.

7. AUTHORS BIOGRAPHY

Kimmo Kokkila holds the current position of Product Manager at ABB Marine & Ports. He is responsible for Azipod C and D products. His expertise includes propulsion ice classes, hydrodynamics and electric power train. He is experienced also in front-end propulsion sales and has worked together with a wide range of customers from a variety of vessel segments.

Jukka Erkintalo holds the current position of Technical Bid Manager at ABB Marine & Ports. He is responsible for yacht and hybrid vessels' power plant and propulsion solutions technical design as well as offering the solutions. His previous experience includes engineering and R&D positions in the Maritime industry together with a strong knowledge of current and future innovative technologies.

APPENDIX 1. Footprint and weight comparison.

Name		DC-POD Onboard dimensions L x W x H [mm]	AC-SHAFT Onboard dimensions L x W x H [mm]	DC-POD Weight [kg]	AC-SHAFT Weight [kg]	Dimensions Result	Weight Result
1) Main Generators		4 pcs 2496 x 1460 x 2013 **14.6 m²**	4 pcs 2496 x 1460 x 2013 1 pcs 2024 x 1250 x 1568 **16.9 m²**	4 pcs 8000 **32000 kg**	4 pcs 7000 1 pcs 4000 **32000 kg**	DC-POD have 3.9 m² less footprint	Equal
2) Low Voltage Switchboards		N/A	1 pcs 4800 x 1050 x 2360 1 pcs 4200 x 1050 x 2360 **9.4 m²**	N/A	5000 4450 **9450 kg**	DC-POD have 9.4 m² less footprint	DC-POD have 9450kg less weight
3) Drives		1 pcs 8730 x 716 x 2088 1 pcs 7530 x 716 x 2088 **11.6 m²**	2pcs 8330 x 716 x 2088 1pcs 3430 x 716 x 2088 **14.4 m²**	1 pcs 6138 1 pcs 5013 **11150 kg**	2 pcs 7400 1 pcs 2100 **16900 kg**	DC-POD have 2.5 m² less footprint	DC-POD have 5750 kg less weight
4) Transformers		2 pcs 1940 x 1375 x 1815 **5.3 m²**	2 pcs 1474 x 1405 x 1760 1 pcs 1700 x 1600 x 1760 **6.9 m²**	2 pcs 3900 **7800 kg**	2 pcs 3150 1 pcs 3500 **9800 kg**	DC-POD have 1.6 m² less footprint	DC-POD have 2000 kg less weight
5) Thruster Motors		1 pcs 1290 x 1000 x 1681 **1.3 m²**	2 pcs 1290 x 1000 x 1681 **2.6 m²**	1 pcs 3300 **3300 kg**	2 pcs 3300 **6600 kg**	DC-POD have 1.3 m² less footprint	DC-POD have 3300 kg less weight
SUBTOTAL of the above fully comparable equipment		**33 m²**	**50 m²**	**54 250 kg**	**74 750 kg**	DC-POD requires 17 m2 less footprint, which is 34 % reduction	DC-POD requires 21 tonnes less weight, which is 27 % reduction
6) Propulsion Motors		N/A	2pcs 4980 x 3035 x 3285 **30.2 m²**	2 pcs 40 000 (Azipod propulsion module w/o propeller) **80 000 kg**	2 pcs 55 000 **110 000 kg**	DC-POD has **30.2 m²** less onboard footprint	DC-POD requires 30 tonnes less weight.
7) Steering Gear (& SRU or rudder)		2 pcs 2500 x 3200 x 1600 **16 m²**	2 pcs, estimated 1610 x 1470 x 875 **4.8 m²**	Not estimated		AC-SHAFT have 11.2 m² less footprint	Not estimated
8) Shaftlines		N/A	2 pcs, estimated 5000 x 700 x 700 **7 m²**	Not estimated		DC-POD have 7 m² less onboard footprint	Not estimated
GRAND TOTAL of items 1-8		**Total 49 m²**	**Total 92 m²**	Not estimated		DC-POD requires **43 m²** less footprint, which is **46 %** reduction	Not estimated

APPENDIX 2. DC-POD single line diagram.

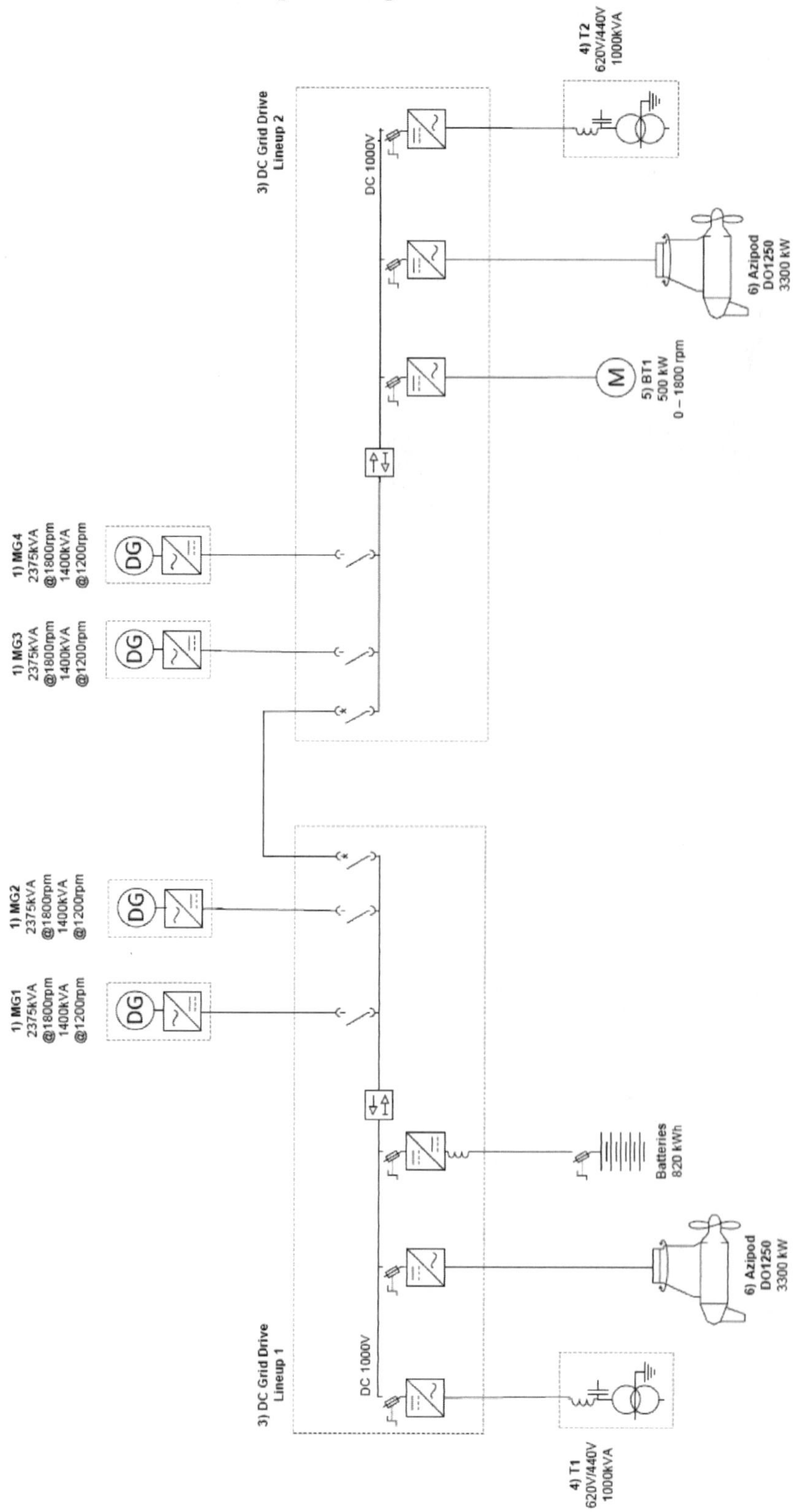

APPENDIX 3. AC-SHAFT single line diagram.

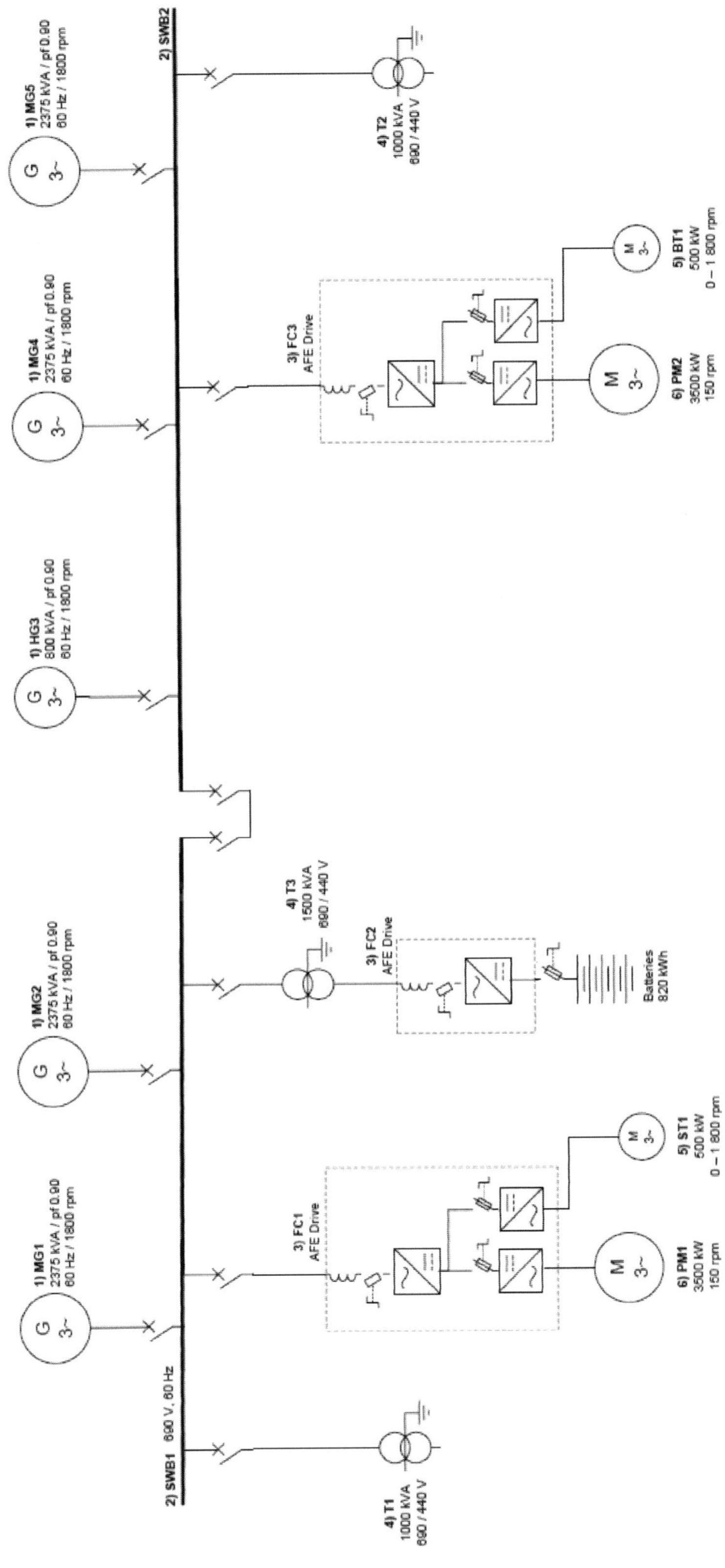

AUTOMATED GENERATION OF DETAILED CABLING DOCUMENTATION FOR SUPER & MEGA YACHTS ACCOMMODATION

Dr. A V Lalechos, LePlan, Greece
D Maketas, LePlan, Switzerland
A Pavlou, LePlan, Greece
I Davaris, LePlan, Greece

SUMMARY

This paper presents a novel approach for the complete design to commission lifecycle of electrical & electronic works of modern, luxury pleasure boats of any size. Smart accommodations are designed and fully documented once technical specifications are defined. Electrical schematics are generated automatically using a set of variable enhanced templates. Templates are selected in an automatic way based on linguistic description of the project. Detailed installation instructions, bill of materials and cable lengths are, among others reports, also generated.

The tool handles among others, low power distribution, lighting, heating/cooling, data networks, voice networks, smart alarm systems, safety systems, access control systems and rack content management.

1. INTRODUCTION

Super and Mega Yacht construction involves the design and cable routing of an increasing number of cables. Introduction of Internet of Things (IoT), smart vessel, enhanced alarm and monitoring systems as well as regular power and control requirements increase cabling requirements and complexity. Currently, conceptual and final plans are difficult to monitor over the complete course of the project phases. This combined with numerous major iterations of plans prior to final implementation created the necessity for an innovative way to handle projects involving above average complexity.

LePlan electrical design and documentation tool is an innovative customised software that converts verbal specifications/concepts using a wording easily understood by naval architects and shipping interior designers into actual electrical plans in an autonomous and automated way. Eventually the tool generates course cable lengths estimates required for budgeting as well as final accurate estimates for procurement purposes.

Additional info supplied are among others, unified electrical drawings for all systems, cable installation instructions, unique cable identifications as well as cable trays and potential penetrations sizing. Finally the resulting drawings are not static and can be utilised by crew during normal operation to timely troubleshoot and repair faults that might occur in the futures

2. CONCEPT

LePlan tool is inspired from modern modular software engineering principles and best practices. Electrical drawings are implemented as a combination of individual single-functions. A subsystem, which is called function, is the combination of several single-functions, which forms a complete system, called group-function is the combination of one or more functions. For simplicity, all levels of functions, irrespectively of their complexity, are commonly called macros.

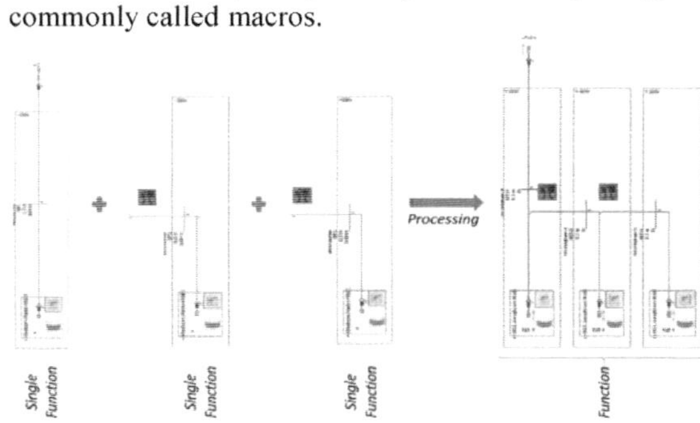

Figure 1: Example of a Function macro built up from three individual single function macros

2.1 MODULAR TEMPLATES/MACROS

The modular design of an electrical system is not something new. Widely available electrical documentation tools have the functionality available for ages; however, in all cases templates remain static.
The term static, refers to templates that are fully predefined. For instance, cables' sizes are pre-fixed and changing the power requirements of a given power line requires the manual alteration in the final drawing, or the adaptation of the template before using it.

LePlan tool goes one-step further and uses a combination of artificial intelligence (AI) scripts that identify which single functions, function or group functions should be used in order to produce the required design and documentation. Therefore, all appropriate and necessary macros are automatically selected, placed on the drawings and required information such as cable's cross-section, cable type, connectors type and size, junction box type, feeding fuses, cables start and end location etc., are automatically populated by the software.

2.2 PROJECT'S SPECIFICATION DATA

Data input for the application originate from a unique source of information, called "Electrical equipment specification form". The data of the form are automatically imported and stored on a local database. Once processed, the data are the only input used by the software in the decision process for sizes, IDs/names, source/destination location etc. The form is mostly drop-down list type to avoid mistype errors and is custom built to fit perfectly each project.

Figure 2: Example of the web-based "Electrical equipment specification form"

An "Electrical equipment specification form" can have any form that can be linked back to a database. For instance, the data can originate from a project's web page, where the client can fill up a dynamic form from a remote location without delaying the specification collection process of the project.

3. DOCUMENTATION GENERATION

Once data collection is verified and initial templates alongside with scripts are designed, the generation of the final documentation is completed within minutes. More importantly, late changes to the data input on the "Electrical equipment specification form" and regeneration of the templates, scripts, and drawings is completed anew within a few minutes.

Making changes can be as simple as reducing low voltage power distribution sockets numbers at a particular location or as substantial as doubling projects size. In a similar way, any closely related project such as a sister vessel can be generated within a matter of hours. Even more significantly, even a non-related project can also be generated within a matter of hours. The need for new single-function templates or not, is the only parameter that can extend the required design time. In most cases no new single functions are needed thus, no new preparatory work is required and once the appropriate form has been submitted, any project can be fully documented within few hours.

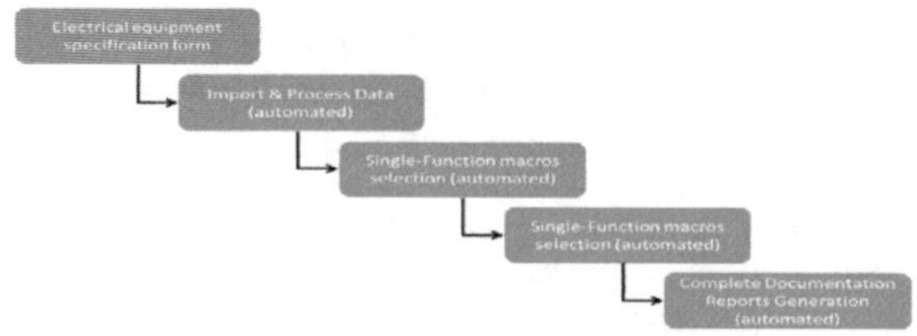

Figure 3: Complete documentation generation flowchart

4. DOCUMENTATION STRUCTURE

The documentation structure follows the ISO/IEC-81346 standard. In principle, a system can be documented using three distinct ways, by answering the simple question "Where I wish to find the documentation," in order to decide how to document a project,

- Location wise:
  When a project is classified by Location, all documents are stored in the source or target location i.e. MSB (main switchboard), EMSB (emergency switchboard), cabinet X etc.

- Product wise
  When a project is classified by product, all documents are typically stored nearby the actual system, i.e. bow-thruster room/cabinet

- Functionally
  In more complex systems, it is common to find documentation is distinct functional folders. A typical example is "Motors List document". Such document holds all the info related to each motor to be found on-board and in most cases includes all/most information related to the automation system handling the operation of each individual motor.

LePlan tool is designed in a way that documentation can be made in all three modes depending on client's needs and preference.

5. NAMING CONVENSIONS

Among the most common problems encountered in most projects, is the use of a non-common naming convention. During the initiation phase of the project, the client and the design team often refer to the same thing with a slightly or even completely different name. Experience and research on these issues, has proved that making changes in later phases of a project involves a huge amount of effort. LePlan tool on the other hand, not only it guides all involved parties into using the same nomenclature but it also allows corrections to be made at any time within a few minutes. All corrections are project-wide and any cross-reference is automatically verified and updated.

6. CABLE ROUTING

In parallel with any other task, the design team analyses 2D and 3D drawings supplied by the naval architect team. The main goal is to identify cable- runways locations and paths as well as penetrations. Once subject information is inserted on the 2D floor plans, they are imported in the tool. This task is time consuming factored to the size of the project and can be as detailed as the cable lengths accuracy needed. When only a course estimate is needed, this task can be typically completed within one or two days in most of the cases.

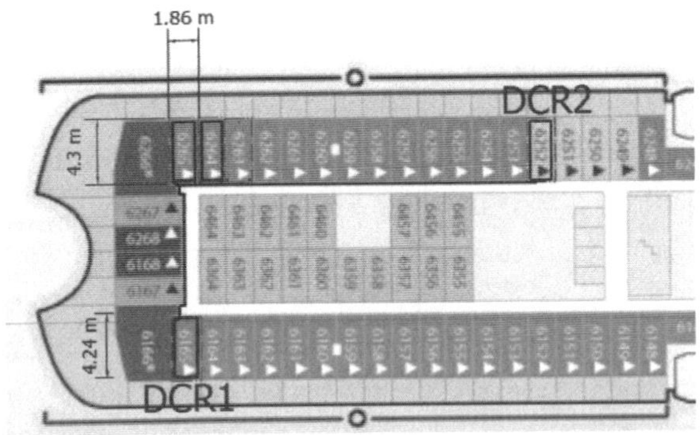

Figure 4: Example of floor plan including potential locations, for two datacentre rooms and one main circular cables runway. Drawing is in 1:1 scale and includes numbered room names.

Once the above step is completed, the manual interaction of the designer and the tool is over. Any cable length can be estimated within few seconds. Changes in the "Electrical equipment specification form" are incorporated in the calculations, almost real-time without any effort from the designer. Just click "update" and wait few seconds. Moving a cabinet to a different deck or room, automatically updates all associated cables. Several reports are available to the design, procurement, and installation team respectively. All reports can be either detailed or generic, with imagination being the only limit.

Cable overview

Cable name	Source (from)	Target (to)
==Distr=220V-WF1	==Distr=220V++OG1.WC+W	==Distr=220V++OG1.HCR-F6
==Distr=220V-WF2	==Distr=220V++OG1.Kitchen+Wall	==Distr=220V++OG1.HCR-F7
==Distr=220V-WF3	==Distr=220V++OG1.Kitchen+Wall	==Distr=220V++OG1.Kitchen+

Figure 5: Cable Overview list including unique cable ID, source, destination, cable type, etc.

7. DEVICES OVERVIEW

One of the biggest problems when budgeting a new project is over-sizing common items. Especially with IT equipment, it is common practice to plan for size one and procure double size. However, this is no longer required. Taking as an example a simple network patch-panel, the ports usage is monitored in real-time. Adding a new device is as simple and as always automated. A free port is always automatically identified by the software and utilised. Additionally via the overview page, it is always possible to see what devices are connected on subject patch panel.

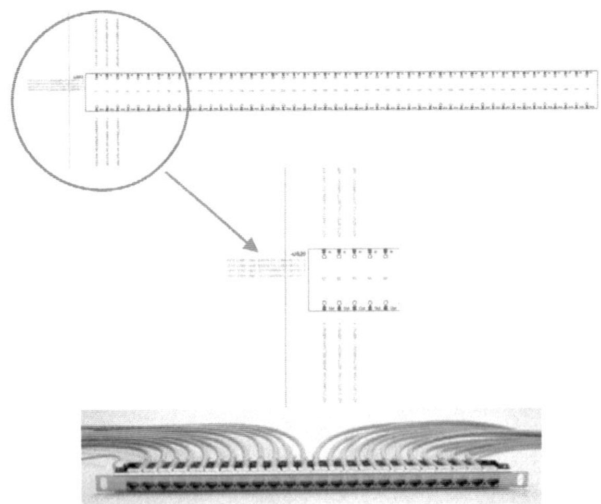

Figure 6: Patch panel ports usage overview

A complete rack view is possible for all common IT equipment or even for custom rack mountable devices. Moving a device from one rack in room X1 deck Z1 to a rack in room X2 on deck Z2 is a trivial task. This gives the opportunity for different vendors to reserve space on racks and reduce the footprint of the equipment used.

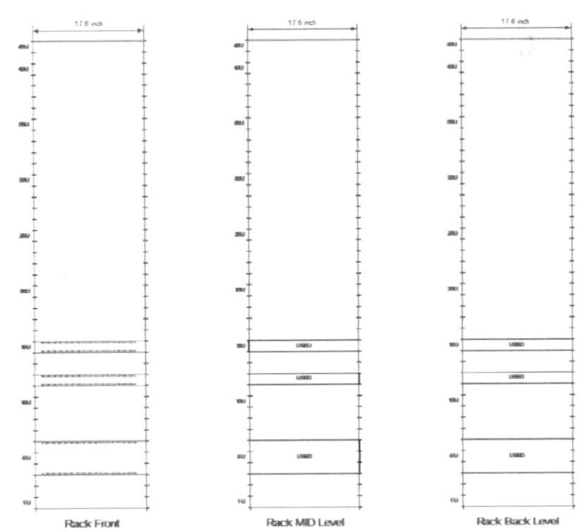

Figure 7: View of a 42U rack including two patch panels and one fully managed network switch[1].

Same principle applies for any component documented using the tool. Among others, these could be a PLC card, an IT device, power and network sockets in a room/area/deck etc. All that is required is to design a function macro for such a device as described in the "concept" section of this paper.

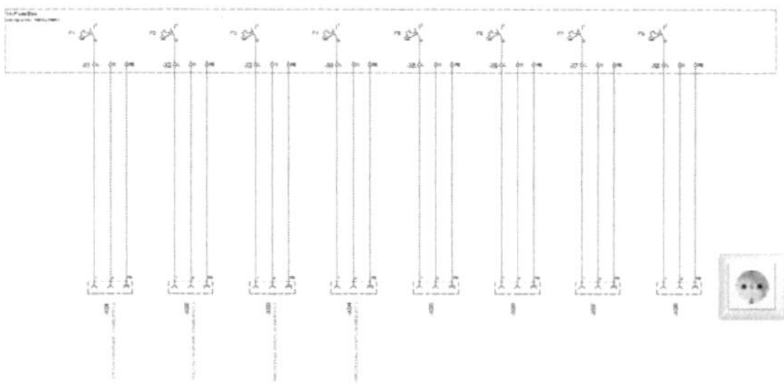

Figure 8: Low voltage power sockets distribution & usage. The green light colour demonstrates that a device is in use and indicates what device is connected on subject socket.

8. CONCLUSIONS

This paper presents a complete unified solution for documenting any project irrespectively of its complexity. Once single function and function macros are designed and implemented, to meet the requirements of a design team or yard, any other project can be documented without extra effort. Repetitive tasks can be incorporated in an effortlessly and quick manner. Detailed or summarised bill of materials are available at any given moment to assist with pre-sales or budget compliance and forecasting. Resulting documentation can be provided to the owner in read only electronic format for easy and accurate troubleshooting in the future.

9. ACKNOWLEDGEMENTS

I would like to express my appreciation to the ePlan Gmbh Company for developing a unique platform serving as the backbone of LePlan tool.

10. REFERENCES

1. Dr. A V Lalechos, 'Design & Documentation at European XFEL Rack Management Concept', 'KITE DESY Engineering and Innovation Day', 2015.

11. AUTHORS BIOGRAPHY

Dr. Antonios-Vassilios Lalechos, holds PhD from University of Southampton, MBA international from University of Portsmouth, MSc in Control system from London Imperial College and BEng in Electrical & Electronic Engineering from University of Portsmouth. Works as Electrical Superintendent while he also holds the position of Leading R&D Consultant at LePlan supporting businesses working with ePlan P8 and EEC One software. Previously he led the Electrical design works for the Experimental Hall of European XFEL project. Among others he developed a tool to unify the documentation of 3.4km tunnels, 6 datacentres and thousands of motors, valves and sensors all part of the same unified design & documentation.

Dimitrios Maketas, holds a MSc in Information Security from Lulea University of Technology in Sweden and a MSc in Software Engineering and BSc (Hons) in Computer Sciences from University of Portsmouth in UK. He has more than 10 years of experience in software and systems engineering with Fortune 500 companies, NASDAQ listed companies and UN agencies. He has co-authored "Formal presentation of fuzzy systems with multiple sensor inputs" with Dr. Alexander Gegov, published in Sensors & Transducers magazine, (S&T e-digest), 08.2005, p. 366-373. He currently works as an IT Analyst for a UN specialised agency.

Antonios Pavlou, holds a MEng in Naval Architecture & Marine Engineering from National Technical University of Athens. Among others, he ranked 3rd at the EU Student Competition 2011, "Visionary Concepts for Ships and Floating Structures – Visions Olympics", organized by WEGEMT and CESA, with the project "E^4-Containership". His dissertation was part of the publication "Energy efficiency parametric design tool in the framework of holistic ship design optimization", Journal of Engineering for the Maritime Environment, August 2011, with authors E.K. Boulougouris, A.D. Papanikolaou and A.C. Pavlou. He currently works as Superintendent Engineer in Shipping Company.

John Davaris, Holds an MSc in Marine Engineering from University of Southampton and a BEng in Mechanical Engineering from Brunel University. Works as a Technical Superintendent in a Shipping Company with extensive experience in Ship repairs and retrofits. Deeply involved in Planned and Condition Based maintenance systems.

LOW DRAG HULL FORM FOR MEDIUM-LARGE YACHTS

E Begovic and C Bertorello, Università di Napoli Federico II, Italy

SUMMARY

A general sustainability of pleasure craft, low fuel consumption and reduced environmental impact are considered important features. Actually such characteristics are generally connected to poor speed, impractical for short time cruising neither attractive for long range passages. The design and optimization of new hull forms suitable for medium-large size motor yachts with low drag and cruising speed adequate to market requests seem very interesting. An experimental research program focused on semi-displacement craft has been considered by MArine HYdrodinamics Research Group of the University of Naples Federico II. This paper reports the experimental results of towing tests relative to a new hard chine hullform up to Froude Number (*Fr*) 0.6. Resistance data in still water for three displacements considering interceptor effect and heave and pitch RAO in regular sea are provided. Hydrodynamic performances are compared with NPL's 100C MODEL showing favourable behaviour.

NOMENCLATURE

B	maximum beam (m)
B_{WL}	beam at water line (m)
C_B	block coefficient
C_F	frictional resistance coefficient according to ITTC 57, $= \dfrac{0.075}{(\log Re - 2)^2}$
C_R	residual resistance coefficient = C_T-C_F
C_T	total resistance coefficient $C_T = \dfrac{R_T}{0.5 \cdot \rho \cdot v^2 \cdot WS}$
Fr	Froude number based on ship length, $Fr = \dfrac{v}{\sqrt{g \cdot L}}$
L_{OA}	length over all (m)
L_{WL}	length at waterline
LCG	longitudinal position of center of gravity, from stern (m)
LCB	longitudinal position of center of buoyancy, from stern (m)
L/B	geometric slenderness ratio
P_B	brake or installed power (kW)
P_E	effective power (kW)
P/Δ	power to displacement ratio (kW/kg)
RAO	Response Amplitude Operator
Re	Reynolds number $Re = \dfrac{v \cdot L}{\upsilon}$
R_F	frictional resistance (N)
R_R	residual resistance (N)
R_T	total resistance in calm water (N)
T	draft (m)
v	ship, model speed (kn, m/s)
VCG	vertical position of center of gravity, from baseline (m)
$L/V^{1/3}$	slenderness ratio
WS	wetted surface, (m^2)
Δ	displacement (kg)
η	total propulsion and hull efficiency
λ	model to ship scale ratio
ρ	water density (kg/dm^3)
τ	running trim angle (deg)
υ	Kinematic viscosity (N s m^{-2})

1. INTRODUCTION

Within the recent general attention to sustainable yachting a peculiar interest has grown up. It is focused on comfortable, low horsepower, long range medium size yachts with adequate speed. Although not original neither new this concept is today considered in a global frame very different from earliest years of this century. Then the excess in performances was the most distinctive character of new designs and low speed displacement yachts were a small niche of the market.

At present, comfort characteristics are most interesting with strong influences on layouts and sea-keeping performances. Moreover, a reduced horsepower is the key factor to get low emissions, long range and sustainable navigation. Such features are peculiar of several proposals of medium and large yacht up to 100' L_{OA}. Unfortunately, they generally present very low cruising speed. Few examples with better maximum speed need large horsepower unacceptable within the mentioned trend. To get an effectively sustainable navigation dedicated and optimized hull forms are necessary. When L_{OA} is over 50 m (160') even *Fr* values below 0.3, peculiar of displacement hull forms, allow adequate cruising speed around 14 kn. At 80' L_{OA}, an acceptable cruising speed over 16 knots means *Fr* around 0.5. This results in a critical semi-displacement hydrodynamic regime characterized by partial hydrodynamic lift and longitudinal trim angle modification. The difficulties of resistance assessment in *Fr* range 0.4-0.6 have led to the general consideration of NPL [1] systematic series as a fundamental reference, although a few modifications of the original forms have been proposed. Nevertheless the high horsepower needed and the dynamic instability experienced by round bilge hull form in semi displacement regime have suggested to go for stronger hydrodynamic lift at higher speed or to choose a full displacement navigation at Fr<0.4. In the last four decades hull designers have concentrated their efforts about planing hull forms both for the large request of higher cruising speed and for the quick and simple available resistance prediction tools.

A further interest in low drag and reduced *WS* hull form comes from the more and more requested hybrid propulsion to allow a zero emission cruising condition allowed in Marine Protected Areas. At present the only practical and relatively cheap zero emission propulsion is fully electric with energy supplied by batteries. This leads to two digits horsepower figures.

2. SCIENTIFIC BACKGROUND

A renowned interest towards semi-displacement hull forms is witnessed by several scientific papers published since 2009 [2,3,4,5]. After some years of small craft research focussed almost only on high speed craft in full hydrodynamic lift, the academic and scientific community has paid attention to more sustainable navigation. Unfortunately the physics of the "critical" hydrodynamic regime in the *Fr* range 0.4-0.6 is difficult to describe in a simple way so that the first fundamental researches on semi displacement and preplanning condition have been strictly experimental [1], [6]. They are still very important in the field. Recently some attempts to exploit the actually available CFD tools have been published [2], [7]. The present researches take benefit from results got for dynamic instability [2], spray rails and hard chine influences [8], [9] and assessment of hull coefficients.

Moreover, the most known scientists in the field of small craft hydrodynamics have recently published on the matter of semi-displacement regime. Savitsky in 2014 [10] identifies the semi-displacement condition as high speed displacement giving useful suggestions for hull geometry. Blount et al. [4] define dynamic instability patterns and highlights the advantages of hard chine. Radojcic et al in 2014 [11] provide a fundamental contribute with "simplex" and "complex" mathematical models for resistance assessment.

Other interesting researches more strictly connected to semi-displacement craft design are reported in [12, 13, 14, 15]. Particular features as air cavities and interceptors are investigated in [16].

3. HULL MAIN FEATURES

A sample of actual design proposals has been considered to define research targets. This preliminary investigation highlights the benefits of lighter displacement but also the need of the development of a hull form optimized for semi-displacement and able to reach *Fr* over 0.55.

To get lowest possible horsepower in *Fr* range 0.4-0.6 the trend in resistance increase has to be almost quadratic, with very small wave pattern. This means that frictional and wave resistance should be of the same order of magnitude up to the highest possible *Fr* value.

From scientific background and design references the main hull features influencing wave resistance components can be identified, they are: $L/V^{1/3}$ ratio, waterline angle at bow and Block Coefficient C_B. Each of them has to be minimized as much as possible. Geometric slenderness ratio L/B has to be the highest possible, although limited by stability and layout issues. The longitudinal position of Center of Buoyancy (*LCB*) is also important and must be a compromise between a more forward value preferable at low *Fr* and an after one requested at higher *Fr*.

The frictional component is very important also and asks for reduced wetted surface. This factor is counteracting with some of the previously mentioned features, mainly with high slenderness ratios, but a reduced wetted surface has been obtained with careful line design of hull ends.
It is evident that such approach will limit allowed displacement and internal volume of the whole hull if compared with existing yachts of the same L_{OA} based on the established NPL series. A lighter displacement can be partly accepted and obtained by low structural weight due to available construction technologies, but reduced accommodation volume consequent to low C_B and to narrow fore body is a tough price to pay to the expected achievements. The influence of these key parameters will be identified by three series of tests with $L/V^{1/3}$ ranging from 6.320 to 5.952.

Allowable full scale and experimental results in semi displacement condition highlight the importance of running trim on motion resistance. The necessity of setting it to optimal value by an adjustable device has been considered and the hull model has been fitted with twin interceptors protruding 0, 3 and 6 mm from transom. Hard chine has been considered preferable for better dynamic stability and better resistance behaviour at higher *Fr* as reported in previous work by the Authors [8]. Moreover, hard chine and almost completely flat developable surfaces seems very attractive for yachts that are mainly built in light alloy.

A skeg is generally necessary for hull forms presenting low aft deadrise angles and will fit the hull. It provides adequate course stability, reduces drift when manoeuvring at low speed and yaw in quartering sea.

The designed hull form is shown by Fig.4. A comparison with NPL [1] model 100C that presents very similar non-dimensional ratios is shown in Fig.5 and 6.

The values of main geometrical characteristics for the tested model in the three conditions are reported in the following Table 1.

Table1. Model Main Characteristics

	Model
L_{OA} (m)	2.000
L_{WL} (m)	1.848-1.898-1.947
B (m)	0.539
B_{WL} (m)	0.478-0.492-0.494
T (m)	0.099-0.108-0.116
Δ (kg)	25-30-35
LCG (m)	1.005-1.007-0.995
WS (m^2)	0.645-0.709-0.767
LCB* (m)	0.945 -0.956-0.960
$L/V^{1/3}$	6.320 – 6.108 – 5.952
L/B	3.930
C_B	0.285-0.298-0.314

** from transom

Fig.4a Designed hull form, perspective

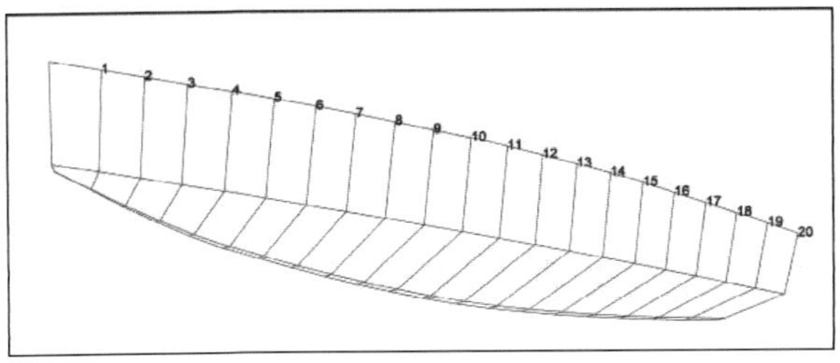

Fig.4b Designed hull form, transversal sections

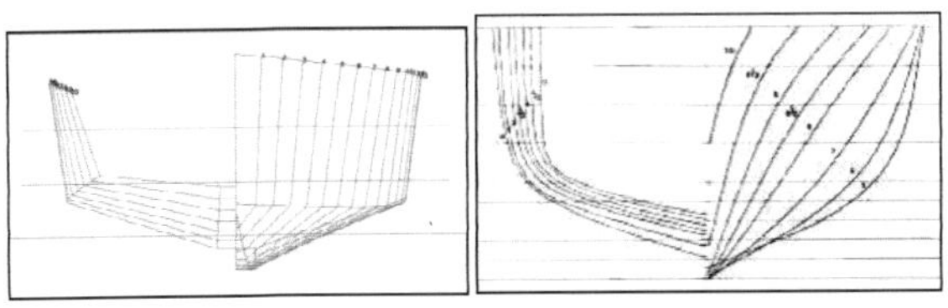

Fig.5 Body plan comparison with NPL100C

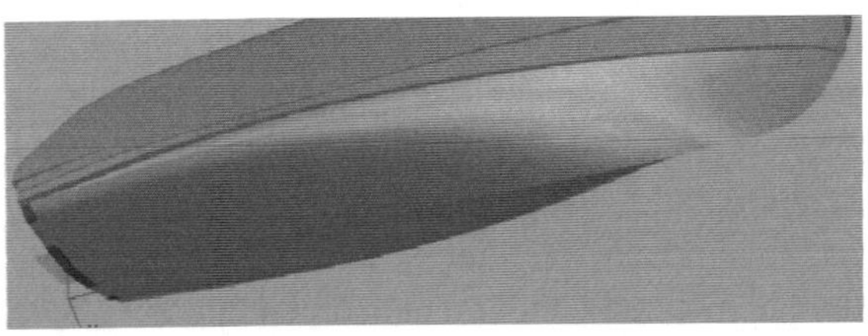

Fig.6 NPL100C Hull Form

4. EXPERIMENTAL TESTS

A 2.00m L_{OA} wooden model of the designed hull has been tested at three displacements 25, 30 and 35 kg with $L/V^{1/3}$ ranging from 6.320 to 5.952. Each condition has been tested in Fr range 0.15-0.70 with model speed ranging from 0.659 to 3.077 m/s. Longitudinal trim at rest has been set to zero for all tests. The towing point was at LCG, vertically positioned close to the VCG. The model was restrained to roll, sway and yaw, and was free to heave and pitch. The measurements of resistance, dynamic sinkage and trim in calm water were performed with the same instrument (metallic arm visible in Figg.7 and 8) at the sampling rate of 500 Hz. The weight of the instrument was part of the model ballast and was also considered in the inertia determination. The tests were performed at Towing tank of University of Naples Federico II, which dimensions are 135 x 9 x 4.2 m with 8 m/s maximum carriage speed.

Fig.7 Towing test in calm water at Fr 0.4 (Ship speed 12 kn)

The model scale ratio λ=12 is relative to a 24 m L_{OA} full scale ship and has been chosen according to towing tank facilities and characteristics. All tests were performed with turbulence simulation provided by a 0.8 mm copper wire set at 5% L_{WL} from the bow.

To evaluate possible dynamic transversal instability towing tests have been repeated with model towed by a wire connected to a load cell and free to roll as shown in Fig.8. Up to the investigated Fr 0.7 the model resulted quite stable and these tests were used to check repeatability and uncertainty of measurements also.

Fig.8 Towing test in calm water at Fr 0.5 with model free to roll

Wave pattern at fore body, shown in Fig. 9 suggests a future application of a spray-rail at bow to reduce the wetted part of the hull sides.

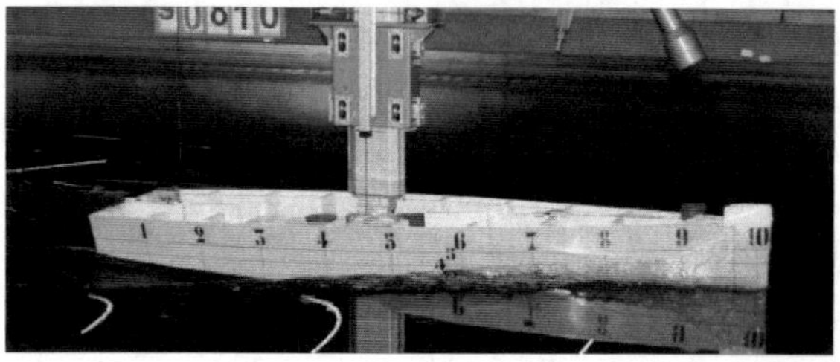

Fig. 9 Towing test in calm water at Fr 0.6 (Ship speed 18 kn)

The model at each displacement has been tested also with interceptors (Fig. 10) protruding 3 and 6 mm from the transom. The 3 mm interceptors, that means 36mm at full scale resulted the best configuration and all the following resistance diagrams are relative to such condition.

Fig.10. Model set up aft view with interceptors

5. RESULT EVALUATION AND COMPARISON

Model Total Resistance (R_T), for the three investigated displacements (25, 30 and 35 kg) is reported in Fig. 11. Displacement influence appears negligible up to Fr 0.4 (12 kn speed for a 24 L_{OA} m ship), but significant at Fr 0.55 (16.2 kn ship speed). The general resistance trend highlights the importance of reducing displacement as much as possible and to consider structural and fitting out weights in the most careful way to get even better performances using lower $L/V^{1/3}$ values. Fr 0.4 represents the most favourable hydrodynamic condition both for very low resistance and poor displacement influence. This value is significantly high and represent one of the achievements of this project.

Total, frictional and residual resistance coefficients (C_T, C_F, C_R) relative to model set at 35 kg displacement are reported in Fig 12. Resistance breakdown confirms previous observations and highlights the same value for frictional and residual components in the vicinity of Fr =0.4, while the same condition for NPL hull form is presented at quite lower Fr below 0.3. This is a very interesting result that confirms a reduced wave pattern for the presented hull form. The trends of residual and total resistance coefficients show favourable behaviour also at high Fr and suggest the exploitation of this hull form at higher cruising speed also.

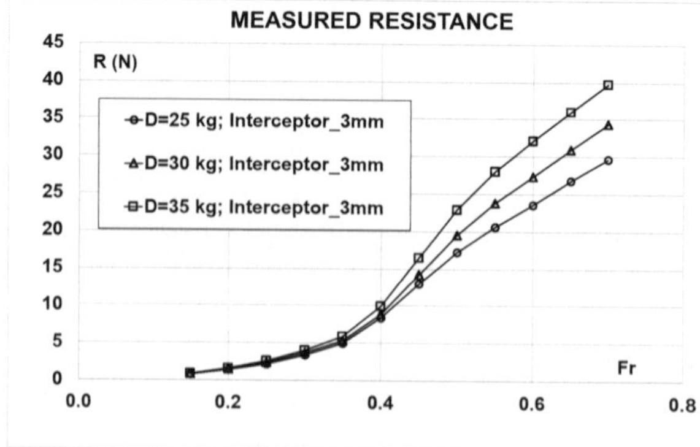

Fig.11 Model Total Resistance, for the three investigated displacements (25, 30 and 35 kg)

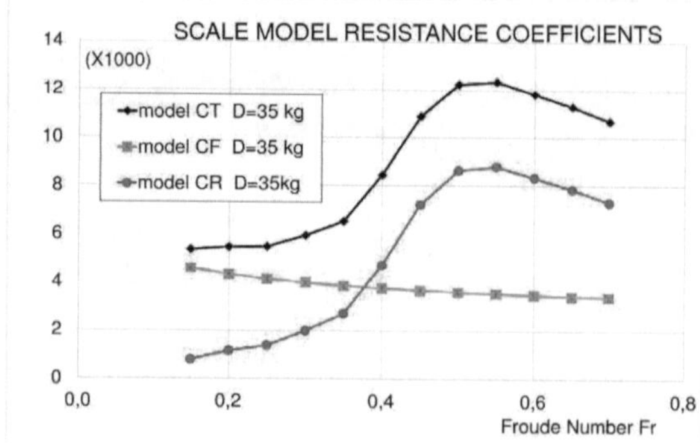

Fig.12 Total, frictional and residual resistance coefficients relative to model set at 35 kg displacement

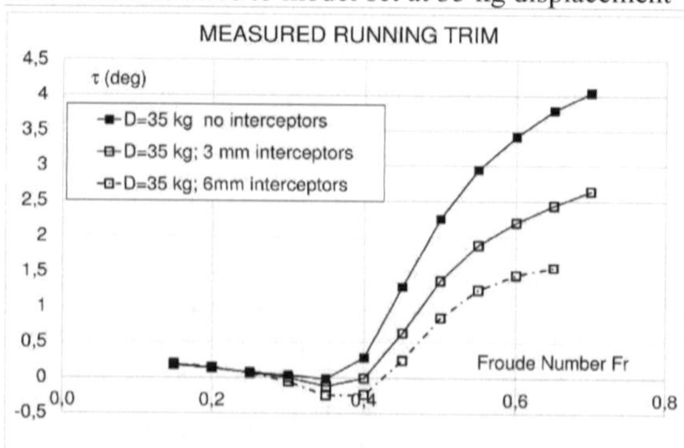

Fig.13 Running trim values of model set at 35 kg displacement for different interceptor extensions

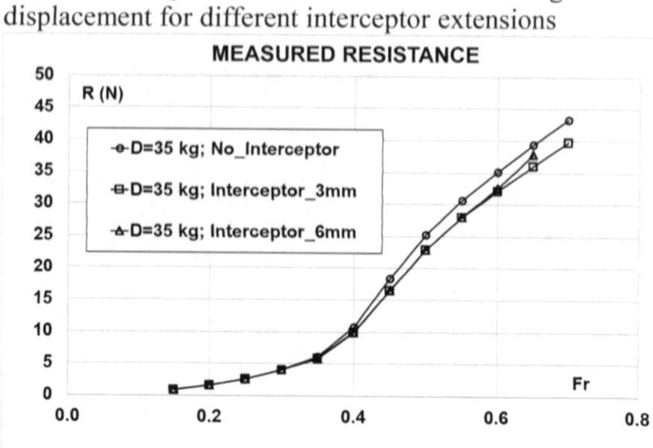

Fig.14 Scale Model (35 kg Δ) measured resistance with interceptors protruding 0, 3 and 6 mm from transom

Running trim (τ) angles of model set at 35 kg displacement for different interceptor extensions are reported in Fig.13. Total resistance values relative to the same model set up and different interceptor extensions are reported in Fig. 14. It is possible to appreciate the benefits of 3 mm interceptors resulting in 1.5 running trim reduction at *Fr* 0.7. Deeper interceptors of 6 mm resulted in too small trim angles at higher speeds and test were stopped at *Fr* 0.65. The interesting feature of the tested interceptors (36 mm deep at full scale) is their fixed position for any speed that does not need any powered or mechanical device as no negative effect is observed at lowest speed values.

Fig.15 reports the Effective Power values for a 24 m L_{OA} ship based on the present hull form and compares them with those ones relative to NPL 100C model reported at the same displacement. A significantly better behaviour can be observed in the whole *Fr* range. The shown differences of Effective Power (P_E) in the order of 20% or more mean differences in Brake power (P_B) in the order of 40% due to usual total efficiency (η value around 0,55. ($P_B = P_E / \eta$).

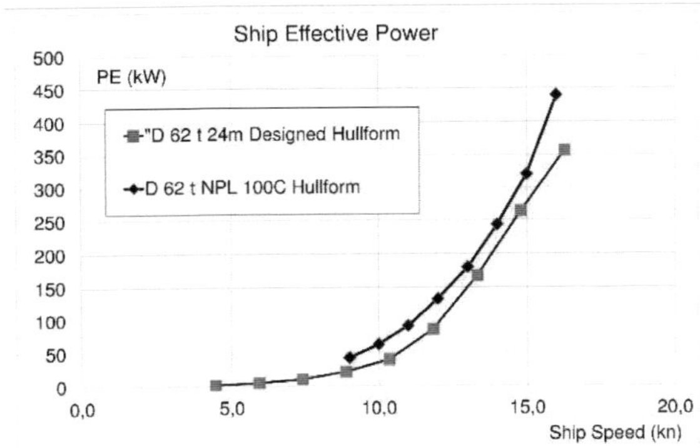

Fig.15. Designed Hull Form and NPL100C Effective Power comparison in full scale

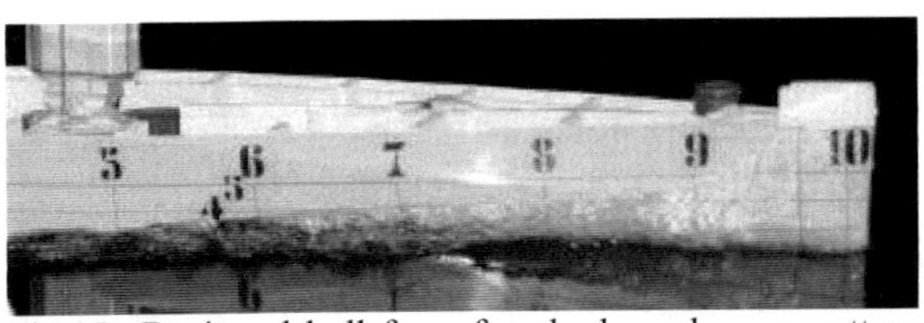

Fig.15a Designed hull form fore body and wave pattern at *Fr* 0.55

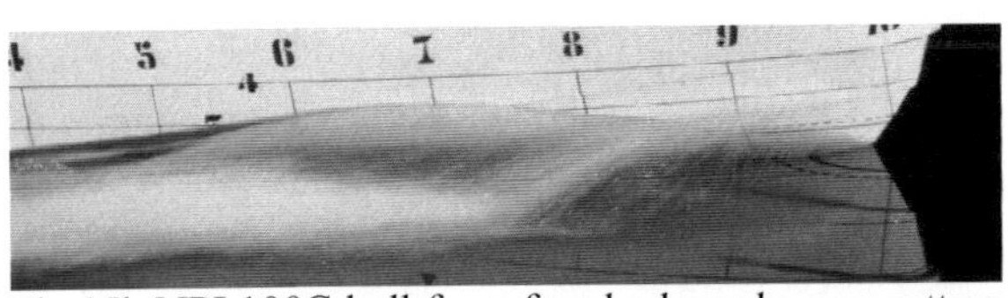

Fig.15b NPL100C hull form fore body and wave pattern at *Fr* 0.55

Figures 15a and 15b show the differences in fore body and waterline angle at bow between the designed hullform and NPL series. This last presents a quite larger wave pattern and spray, which is, at least partly, responsible of the higher residual resistance.

6. SEAKEEPING TESTS IN REGULAR WAVES

The scale model at displacement of 35 kg with 3 mm interceptors has been tested in regular waves at Fr=0.2 and 0.4. Experimental setup for the seakeeping tests is the same as shown in Figs. 7 and 9. Tested waves and respective amplitudes are given in Table 2, together with corresponding wave periods for ship-model scale 12. The metallic arm is able to measure resistance, vertical displacement and pitch at the towed position, that is 0.875m from the transom. The heave response at *CG* is obtained from vertical motion decomposition on heave and pitch. Sampling acquisition is 500 Hz. Preliminary obtained results are presented in Figs. 16 and 17, in standard non-dimensional form of *RAO*, i.e. heave amplitude η_3 is divided by wave amplitude *A*; pitch amplitude η_5 in radians is divided by wave number *k* multiplied by wave amplitude *A*. Further seakeeping tests in both regular and irregular waves are in progress.

Table 2. Regular waves set up

T_{WAVE}	ω	λ	λ/L	A	H/λ
(s)	(rad/s)	(m)	-	(mm)	-
2.89	7.540	1.084	0.550	13.7	1/75
3.15	6.912	1.290	0.655	19.0	1/67
3.46	6.283	1.561	0.792	19.8	1/75
3.85	5.655	1.927	0.978	20.9	1/100
4.33	5.027	2.439	1.238	24.9	1/100
4.95	4.398	3.185	1.617	32.3	1/100
5.77	3.770	4.335	2.201	43.3	1/100
6.93	3.142	6.243	3.169	59.4	1/100
8.66	2.513	9.755	4.952	58.1	1/167
11.55	1.885	17.342	8.803	60.5	1/333

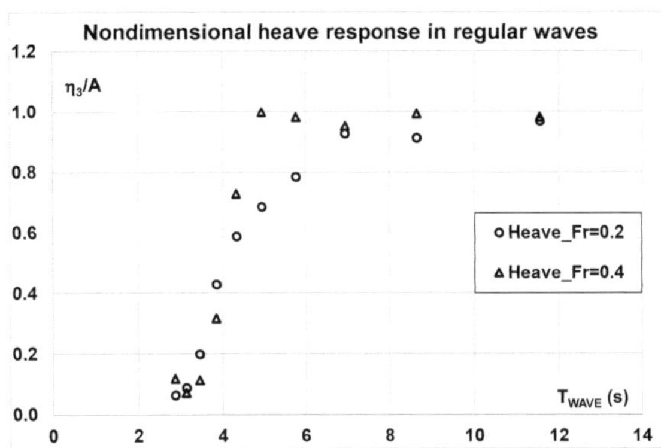

Fig.16. Heave response at Fr = 0.2 and 0.4 in regular waves

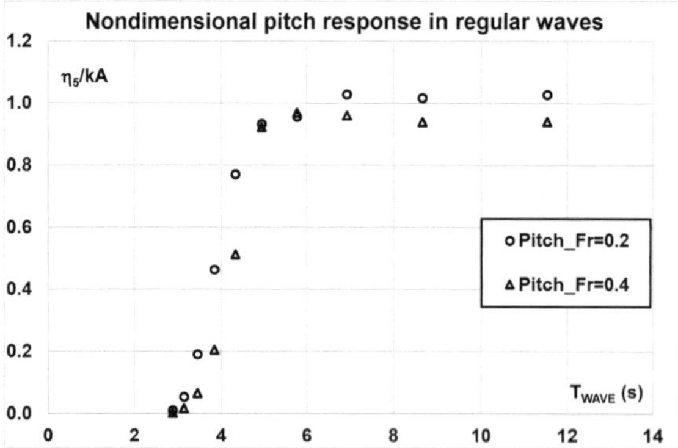

Fig.17. Pitch response at Fr = 0.2 and 0.4 in regular waves

7. IN PROGRESS DEVELOPMENTS AND CONCLUSIONS

The Resistance and seakeeping in regular waves assessment are only the first achievement of a new hull form design. As first results have been promising both for resistance values and dynamic transversal stability, furthers seakeeping tests (Fig. 18) are currently in progress. Their results will be presented in the next future. Moreover two new models with L/B ratio 3.5 and 4.5 respectively have been commissioned and will be tested at the same displacement values to make available systematic data on this promising hull form.

Fig.18 Towing test in irregular wave at Fr 0.6 (Ship speed 18 kn)

8. ACKNOWLEDGEMENTS

This work has been financially supported by University of Naples "Federico II" within the frame of the 2015-2016 research program.

9. REFERENCES

References should be numbered in the text where they occur, with the number in square brackets [1] and a complete list given at the end of the paper.

1. BAILEY D., 'The NPL Round Bilge Displacement Hull Series', *Maritime Technology Monograph No. 4, Royal institution of naval Architects,* 1976.

2. VAN OOSSANEN P., HEIMANN J., HENRICHS J., HOCKIRCH K., 'Motor Yacht Hull Form Design For The Displacement To Semi Displacement Speed Range', *FAST 2009, Athens, Greece pp. 629-643*, October 2009.

3. RANZENBACH R., BOWLES J., 'Hull Forms With Low Cruise Speed Resistance And High Speed Dynamic Stability – Experiments And Computations', *The 29th American Towing Tank Conference Annapolis Maryland*, August 2010.

4. BLOUNT D.L., MC GRATH J.A., 'Resistance Characteristics Of Semi-Displacement Mega Yacht Hull Forms', *Trans RINA, Vol. 151, Part B2, Intl J Small Craft tech, Jul-Dec*, 2009.

5. BRAY P.J., 'Super Fuel Efficient Long Range Motoryachts', *SuperYacht Symposium SNAME*, 2006.

6. COMPTON, R.H., 'Resistance Of A Systematic Series Of Semi Planing Transom Stern Hulls', *Marine Technology, Vol. 23, No. 4, pp. 345-370,* 1986.

7. VAN OOSSANEN P., 'Don't Get The Hump', *The Yacht Report issue 114 pp. 106-109.*

8. BEGOVIC E., BERTORELLO C., BOVE A., PENNINO S., 'Hydrodynamic Performances Of Warped Hard Chine Displacement Hull Form', *Trans RINA, Vol. 156, Part B, International Journal Small Craft Technology, Jan-Jun 05-26,* 2014.

9. BEGOVIC E., BERTORELLO C., PRPIC-ORSIC J., *ROLL DAMPING COEFFICIENTS ASSESSMENT AND COMPARISON FOR ROUND BILGE AND HARD CHINE HULLFORMS,* Proceedings of the ASME 2013 32nd International Conference on Ocean, Offshore and Arctic Engineering OMAE2013, June 9-14, Nantes, France, 2013.

10. SAVITSKY D., 'Semidisplacements Hulls A Misnomer?', Fourth SNAME Chesapeake Powered *Symposium June 23, 2, Annapolis Maryland,* 2014.

11. RADOJCIC D., ZGRADIC A., KALAJDZIC M., SIMIC A., 'Resistance Predictions For Hard Chine Hulls In The Pre-Planing Regime', *Polish Maritime Research 2(82) Vol 21 pp. 9-26*, 2014.

12. TARAVELLA B., 'A Hybrid Method For Predicting Lift And Drag Od Semi-Planing /Semidisplacement Hull Forms', *University of new Orleans Theses and Dissertations, Paper 987,* 2009.

13. RADOJCIC D., PRINCEVAC M., RODIC T., 'Resistance And Trim Predictions For The Sklad Semi-Displacement Hull Series', *Ocean Engineering International Vol. 3, No.1, pp.35-40*, 1999.

14. RADOJCIC D., GRIGOROPOULOS G.J., RODIC T., KUVELIC T., DAMALA D.P., 'The Resistance And Trim Of Semidisplacement, Double Chine, Transom Stern Hulls', *Proceedings of Sixth International Conference on Fast Sea Transportation FAST 2001 Southampton, pp.187-195, September* 2001.

15. GRIGOROPOULOS G.J., LOUKAKIS T.A., 'Resistance And Seakeeping Characteristics Of A Systematic Series In The Pre-Planing Condition', (Part1), *Transactions of the Society of Naval Architects and Marine Engineers, Vol. 110, pp. 77-113*, 2002.

16. MATVEEV K.I., 'Hydrodynamic Modeling Of Semiplaning Hulls With Air Cavities', *International Journal of Naval Architecture and Ocean Engineering Vol. 7 Issue 3 May, pp. 500-508,* 2015.

10. AUTHORS BIOGRAPHY

Ermina Begovic Naval Architect, Ph. D., graduated from University of Zagreb in 1994. She took her Ph.D. at University of Naples Federico II in 2003 with the Thesis: "Cross Flow Effects in Vertical Motion Prediction of Fast Slender Hulls". She is researcher at University of Naples Federico II from 2005 where she is actually Professor of "Ship Seakeeping". Research interest are experimental and numerical assessment of resistance and seakeeping of high speed semidisplacement ships and planing craft, seakeeping performance of damaged ships and added resistance in waves.

Carlo Bertorello Naval Architect, Ph. D., Researcher at University of Naples Federico II from 1999 where he is actually Professor of Ship Design. Professor of Naval Architecture and Marine Construction at Master Course of Yacht Design at Politecnico di Milano from 2001. Member of Italian Delegation at IMO SLF Sessions in 2000 and 2003. Author of scientific papers on: HSC hullforms and performance optimization, multihull ship stability, multiattribute design procedures, composite materials. At present involved in research programs concerning non-monohedral planing hull forms, experimental assessment and optimization of HSC seakeeping characteristics, aerodynamic resistance of HSC.

MULDER THIRTYSIX
SPECIFICATIONS

LENGTH	**36.00 M**
BEAM	**8.00 M**
DRAFT	**2.05 M**
DISPLACEMENT	**190 TON**
GROSS TONNAGE	**298 GT**
TOP SPEED	**16.8 KNOTS**
DIESEL	**32.000 L**
WATER	**6.000 L**
MATERIAL	**ALUMINIUM**
CABINS	**FIVE LUXURIOUS STATEROOMS**
ENGINES	**CATERPILLAR C18/1150HP**
HULL SHAPE	**FAST DISPLACEMENT HULL FORM (FDHF)**
CERTIFICATION	**MCA + LY3**
CREW	**4-6 PERSONS**
NAVAL ARCHITECT	**VAN OOSSANEN NAVAL ARCHITECTS**
DESIGN	**CLAYDON REEVES DESIGN STUDIO**
INTERIOR DESIGN	**CUSTOM BUILD**
TENDER	**WILLIAMS 625 DIESELJET**
TOYS	**2X SEADOO LUXURY 3-SEATER**

SCPH2: A NEW TECHNOLOGY FOR LOWER RESISTANCE DEEP-V PLANING HULLS

S Brizzolara, Kevin T. Crofton Department of Aerospace and Ocean Engineering, Virginia Tech, USA

SUMMARY

A new hull technology that has potential to considerably improve the performance of a parent deep-V hull is presented. We named this hull technology: Stepped Cambered Planing Hull with Hydrofoils, $SCPH^2$. Inspired by the Dynaplane concept introduced by Clement and Koelbel (1993), the $SCPH^2$ has been conceived for its application to Deep-V hulls (deadrise angle > 20 degrees at stern) and augmented with new super-cavitating hydrofoils at stern and new design of the shape of the swept back cambered planing surface. The main theoretical elements that justify the superior performance of a Deep-V SCPH2 hull with Hydrofoils with respect to a conventional deep-V equivalent planing hull are laid down in the paper. The demonstration of the hydrodynamic advantages will be given on a practical design case for which comparative results of towing tank tests on a modified design of the hull 5631 of the USCG systematic series (Metcalf et. Al. 2005) of contemporary deep-V planing hulls. High-fidelity viscous fully non-linear free surface numerical fluid dynamic simulations will help the discussion. Results obtained from high quality model tests (with underwater photography) allows to validate the main assumptions of the theoretical design method and the CFD predictions and confirm the the physical sources that cause a measured resistance reduction of more than 30% (in model scale).

NOMENCLATURE

$A.R.$	Aspect Ratio ($b/l_m=1/\lambda$)
B_{OA}	Breadth over all [m]
β, β_T	Dead-rise angle at midship, at transom [deg]
b	Maximum breadth at chine [m]
$C_{L,D}$	Lift or Drag Coefficient $2(L, D)/(\rho b V^2)$
D	Drag force [kg]
γ	Spray Root Angle [deg]
∇	Displacement volume (at rest) [m^3]
Δ	Vessel Displacement [kg]
Fr_∇	Volumetric Froude number ($V g^{-1/2} \nabla^{-1/6}$)
ϕ	Half chord line angle [deg]
L_{OA}	Vessel length over all [m]
L_{WL}	Vessel waterline length [m]
L_{CG}	Longitudinal centre of gravity [m]
L_{CB}	Longitudinal centre of buoyancy [m]
V_{CG}	Vertical centre of gravity [m]
L_{CP}	Longitudinal centre of pressure (dynamic) [m]
$L_{k,c}$	Keel / Chine wetted length [m]
l_m	Mean chord length $(l_k+l_c)/2$ [m]
ρ	Water mass density [kg m^{-3}]
S	Hull Pressure Area [m^2]
$\tau_{S,D}$	Static/Dynamic trim angle [deg, positive bow-up]
V	Relative Inflow speed [m/s]
v	Vessel speed [knots]
VoF	Volume of fluid

1. INTRODUCTION

Since from the '60s, when they were first introduced, Deep-V planing hull forms have taken the lead of the most popular planing hull type for high speed yachts. Their inherent superior seakeeping performance have determined their success and side diffusion, but they are known to be affected by a considerably higher resistance to the advance motion. For many (e.g. Acapora, 1995) deep-V hull forms remain the biggest innovation in planing hull design up to date.

One of the principal research goals of the innovative Ship design laboratory, at MIT first and now at Virginia Tech, is to improve the hydrodynamic characteristics of deep-V hull forms, by introducing new design elements and fluid dynamic features.

Studies started years ago (Brizzolara and Federici, 2010) with the early fully turbulent viscous two phase simulations of partially stepped hull forms, that for first demonstrated the possibility of accurately predict the hydrodynamic characteristics and complex flow features of this particular type of planing hulls. The same numerical models have been fundamental for the design of a higher efficiency swept-back stepped version of a contemporary deep-V high speed planing craft, that demonstrated a drag reduction of more than 15%

(Brizzolara and Federici, 2013), when compared to the original hull.

In this paper, we describe the main results of a recent research effort aimed to revisit the original Dynaplane concept introduced by Clement and Koelbel (1993) and rethink its application to high deadrise (midship deadrise angle greater than about 20 degrees) planing hulls. We call this planing hull technology Stepped Cambered Planing Hull with Hydrofoils: $SCPH^2$. It is remarkable that many decades after the early visionary research body that Clement managed to generate at the David Taylor Model Basin since the years '60s, few Dynaplane boat designs have been built with success. One obvious reason is that the design is not trivial and there are indeed design issues yet left open after the visionary work of Clement. Another reason is that modern seagoing planing hulls are all deep-V type. In fact, Clement did not consider - actually avoid - the design of highly swept back, cambered stepped planing surfaces with high deadrise, i.e. greater than 15 degrees. The technical justification of this refuse was not given and the application domain was simply restricted to rather flat bottom planing hulls that unfortunately do not meet application interest for current high speed Naval crafts, nor for fast seagoing yachts.

The next sections show first the fundamental principles at the base of the higher efficiency expected from the $SCPH^2$ new hull typology. Then we introduce and discuss, with the aid of high-fidelity viscous non-linear-free-surface flow numerical simulations, a practical design example. Theoretical and numerical predictions are finally confirmed by high quality model tests that highlights several areas of potential further improvement of the technology, under current investigation in a newer research project.

2. FUNDAMENTAL PRINCIPLES OF THE $SCPH^2$

We adopt Clement's lifting surface approach (Clement, 1964) to discuss the hydrodynamic characteristics of planing surfaces. These are the principles that led him towards the Dynaplane stepped planing hull design. Some of these principles are repeated here as they constitute the foundation for the design of the high deadrise $SPCH^2$ presented here. The analogy between lift produced by planing surfaces and thin lifting surfaces developing lift on the pressure side only, leads to the recognition that the aspect ratio must influence the hydrodynamic efficiency, in terms of lift over drag ratio.

Consequently, following classical aerodynamic theory, *High aspect ratio planing surfaces are more efficient than low aspect ratio planing surfaces.* The aspect ratio AR of a prismatic V-planing surface is defined as the ratio of the chine breadth over the mean wetted length, hence it is the inverse of the mean wetted length, $AR=1/\lambda$. A typical value for a conventional deep-V planing hull is typically below 0.5, at pure planing speeds ($Fn_\nabla>4.5$). Alike wings, the induced drag of a planing surface decreases rapidly when the aspect ratio increases (at given lift).

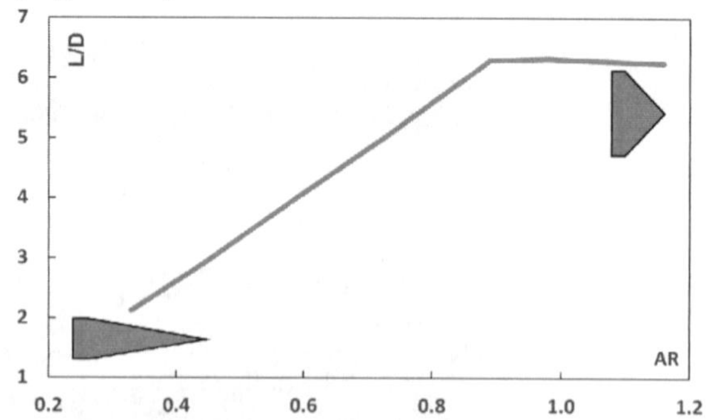

Figure 1: Maximum lift over drag versus aspect ratio for V planing surfaces with deadrise β=25 deg.

Figure 1 presents the maximum lift to drag ratios (i.e. efficiency) calculated according Savitsky method, for prismatic deep-V planing surfaces (β=25 deg) at constant lift. The maximum value of the lift over drag ratio is attained at $AR\sim1$, where it finds a plateau. At these high deadrise angles, high aspect ratios (AR>0.8) can be practically reached only with a single step configuration, i.e. by interrupting the wetted area of the planing surface with a step shortly after the intersection of the spray root line with the chine. A swept back step is preferable in order *to maximize the wetted length at the chine* for a given value of the pressure area and beam at chine. The demonstration is evident by the comparison of the three stepped planing surface presented in Figure 2. All three surfaces have the same wetted area, same aspect ratio and same inclination angle of the spray root line, but have different sweep angle of the step line. Swept back surfaces are able to maximize the chine wetted length (for a given area).

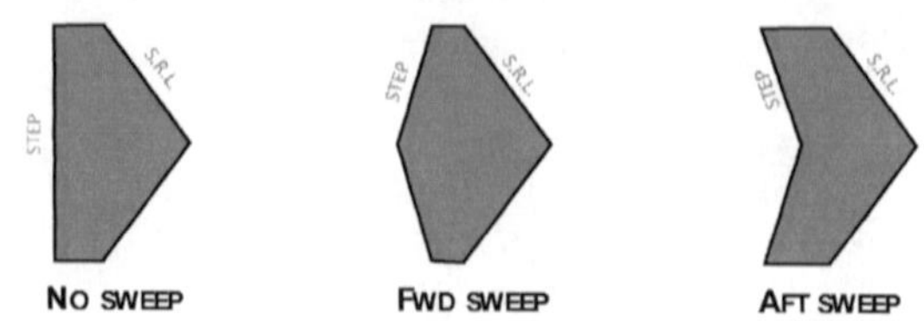

Figure 2: Different step line configurations for planing surfaces of equal area and equal inclination of the spray root line.

A larger wetted length at the chine means a greater margin before the spray root line starts to intersect the step with a change of trim (reduction).

Clearly, the extremely reduced size of the pressure area greatly limits the capability of creating meaningful amount of lift force by just angle of attack. As used in wing aerodynamics and analogously to what is used for super-cavitating hydrofoils, a generous curvature is given to the wetted area bottom of the hull. Hence, another fundamental concept that diversifies $SCPH^2$ from conventional deep-V hulls is the generation of lift by longitudinal camber as opposed to angle of attack. In fact, *for a lifting surface producing lift by camber is more efficient than producing it by angle of attack.*

As a proof, Figure 3 presents the efficiency comparison (in terms of lift over drag) of a flat planing surface versus a cambered planing surface, having the same aspect ratio as a function of the developed lift force. The comparison is based on experimental tests conducted on longitudinally cambered planing surface with no deadrise (Moore,1967) and results of Savitsky method for the equivalent flat plate. In more general cases, as those relevant for the design of $SCPH^2$, a new numerical lifting surface method (Brizzolara & Vernengo, 2016) has been developed that can consider deep-V planing surface with constant of variable deadrise.

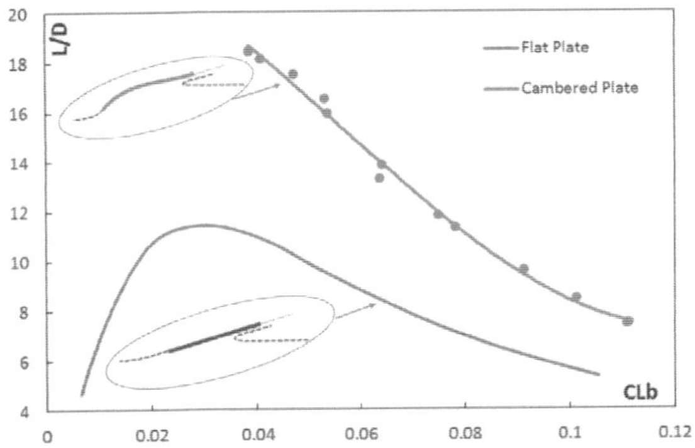

Figure 3: Rectangular Flat (blue) and Cambered (orange) Planing Surfaces, AR=2 and variable trim angle. Comparison of the attainable Lift to Drag ratio vs. Lift Coefficient. The cambered planing surface has a Johnson three term face line.

Finally, the hull after-body (part of the hull aft of the step) is meant to work completely dry in planing conditions. So to guarantee the necessary equilibrium at high speeds, an additional support is needed at stern. We chose to use Surface-Piercing (SP) hydrofoils at stern. As opposed to fully submerged configurations, SP hydrofoils are able to develop a righting trim moment, due to a change of the submerged lifting area. Contrary to Clement simple hydrofoil layout, we use a special design, based on a new SC section (Brizzolara, 2015), that allows them to work in optimal conditions both in fully-wet and in fully-ventilated regimes. This latter regime should be legitimately expected on any SP hydrofoils designed to support planing crafts. Traditional foil sections lose their ability to create lift when super-cavitation (or full-ventilation) occurs (e.g. Brizzolara & Bonfiglio, 2015).

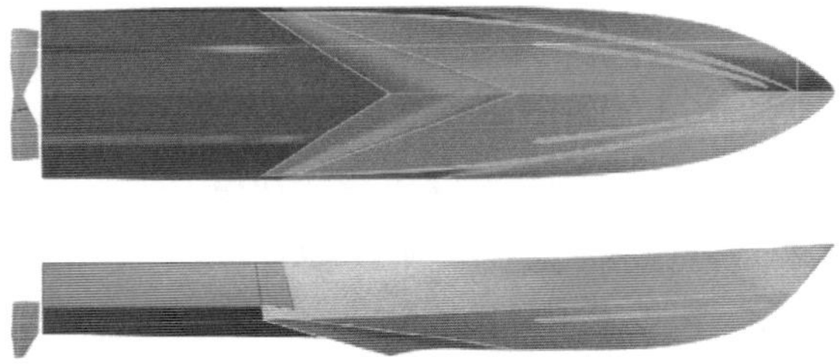

Figure 4: Main Elements of a $SCPH^2$ configuration: fore-body (blue), aft-body (magenta), cambered bottom with step (orange), surface-piercing super-cavitating V-hydrofoils (Green) at stern.

3. DESIGN CASE EXAMPLE

The goal of the research project is to prove that a Stepped Cambered Planing Hull with Hydrofoils ($SPCH^2$) would be more efficient than an equivalent conventional planing hull. To ensure a fair and consistent comparison between the reference and modified hull forms, it was decided to develop the $SPCH^2$ variant of a modern planing hull with good proven characteristics. To this scope, the hull of the model 5631 of the USCG planing hulls mini-series (Metcalf et al., 2005) was chosen. The main characteristics of this hull are given in Table 1, together with the parent hull of the series. The body plan of this hull is represented in Figure 4. This hull was chosen among those of the series because of its higher deadrise: 20 degrees at the transom and about 25 degrees at midship, well in line with modern seagoing planing hull designs that need higher seaworthiness in rough seas. The hull was tested in towing tank at three different displacement values and two different LCG= 38% and 42%.

The reference condition was selected to reflect the intermediate displacement (375lbs, model scale) predicted for the full scale vessel and the load distribution showing the best performance on the reference hull, i.e. LCG=42%. The $SCPH^2$ variant was design and optimized for this reference condition (Faison, 2014), intelligently mixing empirical design procedure with systematic CFD simulations. The design speed was selected to correspond to $Fn_\nabla=5.5$.

Table 1: Main Geometric Characteristics of the reference null (model 5631) and parent model (5628) of the USCG planing hulls mini-series

MODEL	LBP	B	T	L/B	B/T	β	∇	A_P
	[ft]	[ft]	[ft]	[/]	[/]	[deg]	[ft³]	[/]
5628	10	3.08	0.608	3.24	5.08	16.61	8.05	25.88
5631	**10**	**2.24**	**0.51**	**4.47**	**5.08**	**20**	**4.88**	**18.76**

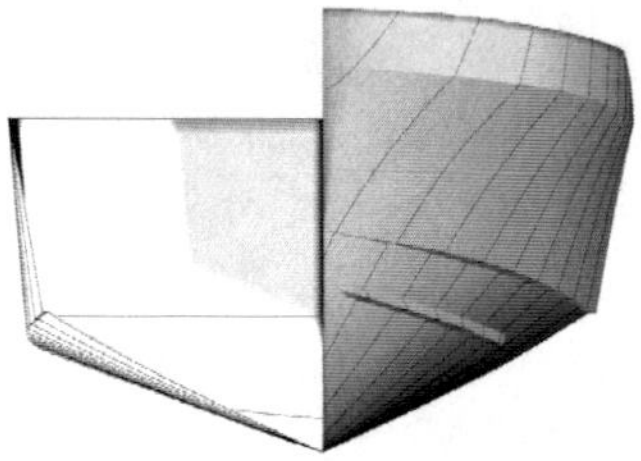

Figure 4: Body Plan of the USCG model 5631

The first problem in the design of the SPCH^2 variant is the selection of the design dynamic trim angle and the identification of the spray-root-line position on the original hull. Due to the additional degree of freedom offered by the two sustaining elements – the cambered bottom forward of the step and the stern hydrofoils – the trim angle can be included among the design free parameters. Its value should be selected in order to get the best efficiency out of the cambered planing surface and Clement proposes a correlation line, whose origin is not better explained.

Since dynamic porpoising stability is of major concern for the design of a SCPH^2, we decided to use a preliminary porpoising stability criteria (Celano, 1998) rigorously valid only for conventional planing hulls, to select the design trim angle τ_{des} = 3.5 deg. A new physics-based porpoising stability criterion, tailored on the specific SCPH2 configuration, was later on devised and verified in course of the design study (Faison, 2014).

The first design problem is to position and size the planform shape of the cambered portion of the bottom in front of the step (yellow area in Table 1.). It is clear that the leading edge if this area should be ideally coincident with the spray root line. A rectilinear SRL is assumed and positioned along the original planing hull, following Savitsky (1964) method, rigorously valid only for prismatic dihedral (V) planing surfaces. To account for the effect of camber, we consider an additional 5 degrees inclination, as suggested by Clement.

Knowing the location of the leading edge line and assuming a given aspect ratio and taper ratio of the cambered planing surface forward the step, the keel and chine chords are calculated.

The aspect ratio AR=2 and taper ratio TR=1:3 were selected according the indications given in Clement (2005). The assumed values seem rather arbitrary and not justified by any particular theoretical argumentation. Some hypothesis behind this uncommented choice could be:

- uncambered dihedral prismatic planing surfaces with deadrise between 5 and 15 degrees reach the best efficiency (L/D) for AR=2 (see Figure 1, for β=10 deg);
- TR=0.35 brings the spanwise lift distribution of non-swept and untwisted trapezoidal lifting surfaces close to the ideal one (elliptic), so it minimizes the induced drag maximizing the efficiency (L/D)

Our design case does not fall within the range of validity of the two previous statements: in fact, the deadrise is well above the upper limit of β=15 with regards to the first above statement and the planing surface is indeed cambered. Moreover the optimum taper ratio is valid for unswept lifting surfaces, while the sweep-back angle is quite large in the current case (mid-chord sweep angle is about 65 deg).

To refine these preliminary and design choices based on approximate theories, a systematic sensitivity analysis able to assess the performance variation around the design point would be recommendable. In this respect, a new medium-fidelity numerical method, called PLAMIT, was developed (Brizzolara & Vernengo, 2016) in parallel to the described design study, but here used only to obtain an alternative cambered planing surface design. A preliminary version of the code, still not fully validated, was used to design an alternative cambered planing surface (version B) with exactly the same planform shape but a different camber distribution along the breadth (optimal as predicted by PLAMIT). The difference in camber between the original camber distribution (A) and the optimal (B) is substantial with less camber at the keel and at the chine in (B) and more camber at midspan with respect to (A). Both versions were built and tested in model scale.

The design of the cambered planing surface was instead verified by a series numerical simulations using a viscous multi-phases (air/water) numerical flow solver and a particular modeling technique first developed and validated in case of different types of stepped planing hulls (Brizzolara & Federici, 2010). Recently, the same CFD models have been used to guide the optimum design of deep-V, swept back stepped planing hulls with good results (Brizzolara & Federici, 2013).

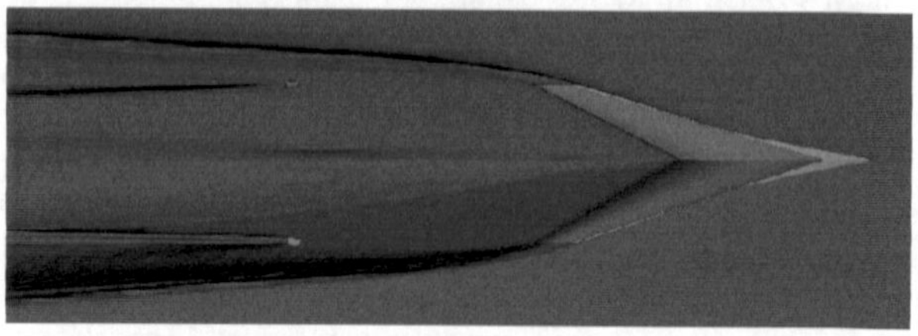

Figure 5: Validation by CFD of the preliminary design of the cambered planing surface foreward of the step. Design planform shape (red) is compared with the actual wetted area of the bottom (green). Predicted free surface represented in blue.

Figure 5 shows the good agreement between the wetted area predicted by the CFD model (green) with the designed area (in red). Simulations were done with the boat running at design speed, free to sink and trim.

The same CFD model was used to guide the design of the afterbody and the stern hydrofoil stabilizers; both components in fact operate in the (free surface) wake washed by the fore-body with cambered bottom in front of the step.

The very next element to be designed is the step height. Based on previous experience in the design of swept back stepped planing hulls, the value of 1% of the beam at the chine suggested by Clement, was overridden with a generous 10% ratio, in order to guarantee the natural ventilation of the aft-body not only at high speed, but also in a wide lower speed range and without venting ducts. The drawback of this choice is the higher resistance at lower speeds, when the ventilation does not occur. This design, though, targeted the best performance at highest speed range only.

The afterbody was then design to remain dry (especially the bottom surface facing the transom) in the free surface wake detaching from the V-step with a sufficient margin of ±2 deg trim, in order to allow for the wake shape modifications at lower speeds and for vertical motions in waves. This after-body design criteria were preliminary implemented assuming a prismatic V-surface detaching from the step edge and extending in the wake as an extrusion of the step edge in the direction of the speed of advance, as represented in Figure 6. Then, once the predicted of the "real" wake detaching from the fore-body has been obtained from the detailed CFD models, the afterbody design is updated. The real free surface wake is generally deeper at stern than the simplified prismatic V-shape wake (yellow planes in Figure 6). It forms a peculiar "W-shape" under the bottom of the transom: a steep wave crest (rooster tail) is visible on the centerline that rapidly slopes down forming a local through on each side (portside and starboard) and then curves up again in a circular arc shape. A section of the calculated free surface for the model at design speed (Fn_{∇}=5.5) is given in Figure 11. This type of wake is well presented and analytically interpolated in Savitsky & Morabito (2010), although only in case of prismatic V-planing surfaces.

The detailed characterization of the free-surface wake aft of cambered swept back planing surfaces is a current research topic under a new ONR sponsored research project. An initial CFD study (Faison, 2014) evidenced how the wake behind a cambered planing surface is substantially influenced by the camber distribution along the breadth, resulting in a largely different from the one that could be estimated with the method proposed by Savitsky & Morabito. The difference is due to the deeper position of the step edge induced by the camber and the higher downwash speed induced by the camber. These effects are captured by the comparison of the predicted flow speeds at the edge of the step by CFD in case of the original and cambered fore-bodies, both simulated at the same dynamic attitude.

Figure 6: Design of the after-body and the stern hydrofoil stabilizers (green objects) with respect to the idealized free surface (yellow V-prism) and the real free surface (blue) predicted in the wake of the cambered stepped planing surface.

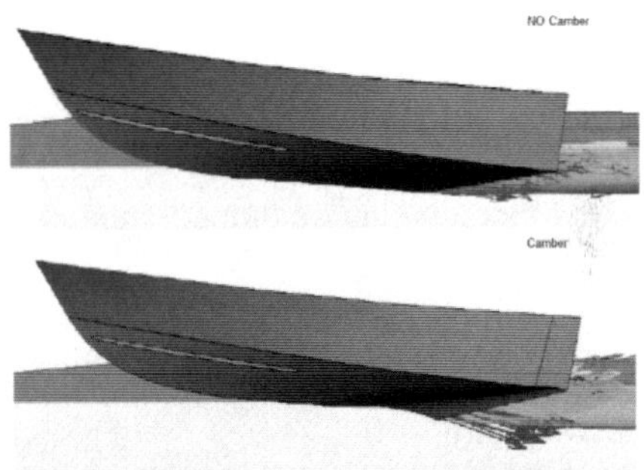

Figure 7: Effect of camber on the predicted free surface shape in the wake of the swept back stepped planing hull. Top picture no camber, lower picture cambered planing surface.

The design the Super-Cavitating Surface-Piercing (SC-SP) hydrofoil stabilizers at stern follows a similar two steps, CFD based, approach as the one used for the aft-body. An earlier research project was aimed to the hydrodynamic design and physical characterization of SC-SP hydrofoils for ultra-high speed hydrofoil crafts. An innovative SC-SP hydrofoil, based on a new family of SC-sections (Brizzolara, 2015), was developed using a non-linear lifting line theory and CFD predictions of the 2D section hydrodynamics. The new hydrofoil, shown in Figure 8, was then tested in a free surface cavitation tunnel confirming an optimal efficiency and stability of the ventilated-flow regime over a wide range of angles of attack, Froude and cavitation numbers.

The same hydrofoil design, opportunely scaled in similitude, is used at the stern of the SCPH^2 hull as pitch stabilizer.

Initially a twin foil arrangement protruding from the sides of the aft-body at stern with a negative dihedral (anhedral) angle was considered, as represented in the 3D views of Figure 9.

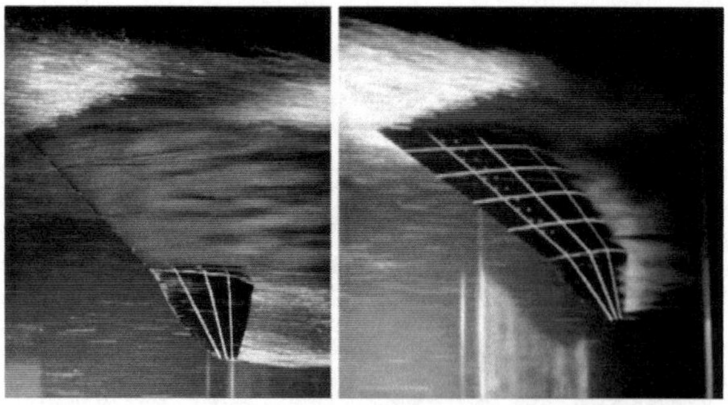

Figure 8: Super-Cavitating, Surface-Piercing Hydrofoil. The pictures capture the back of the hydrofoil in two different regimes: super-cavitating (fully-ventilated) conditions (left) and in non-ventilated condition (right)

The hydrofoil sizing and submergence was chosen in order to support 10% of the vessel displacement while its arm was chosen in order to guarantee the pitch equilibrium, at design speed. After the initial positioning, based on the approximate wake prediction (Figure 6, yellow planes), the length of the hydrofoil had to be increased to allow for the exact shape and deeper location of the computed free surface at stern.

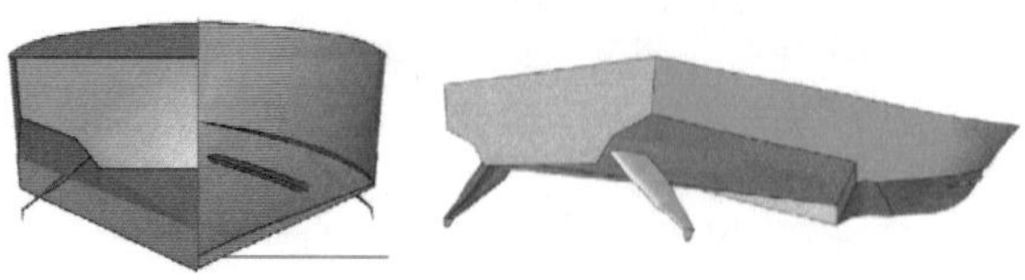

Figure 9: SCPH2 design with Anhedral Surface-Piercing Super-Cavitating Hydrofoil Stabilizers at stern. Body plan view (left) and assonometric view (right).

CFD simulations of the final design configuration (presented in Figure 9) in fixed dynamic attitude confirmed the achievement of the heave and pitch equilibrium, hence confirming the correct sizing and estimation of hydrodynamic forces from the hydrofoils. Unfortunately, further unsteady CFD simulations, with the hull free to heave and sink, evidenced dynamic instability problems of this configuration. Both heave and pitch motions numerically predicted for this first $SCPH^2$ configuration in calm water show a periodic oscillation around the mean (equilibrium) value, as presented in Figure 10. This is a typical motion behavior of a planing hull affected by porpoising instability.

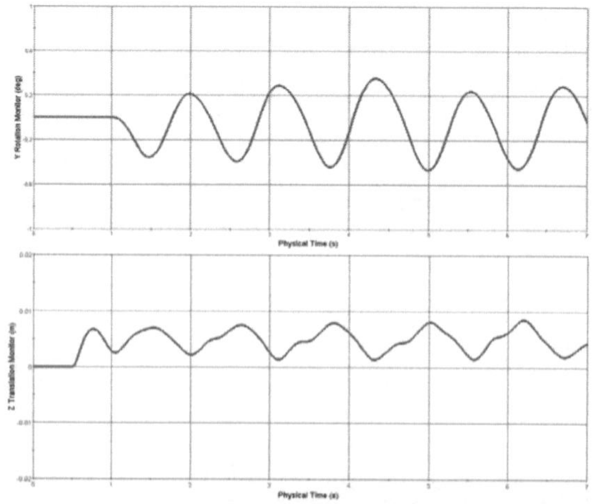

Figure 10: Pitch (top graph) and heave (bottom graph) motions predicted for the SCPH2 design with anhedral stern hydrofoils at design speed, in calm water. Steady State Fluctuations around an average value highlight porpoising instability problems.

It is beyond the scope of this paper to discuss the studies and investigation that were made to better understand and characterize the physical nature of this type of dynamic instability which is extremely important for the successful design of a $SCPH^2$ craft. The result of these studies brought to the definition of a CFD-based methodology to predict porpoising instability of this particular hull configuration and in simplified criteria to assess this quality in a preliminary design stage.

From a practical design standpoint, the ultimate outcome of the porpoising stability study was a change in the SC-SP stern hydrofoil configuration: from twin anhedral side hydrofoils to a centered V-shape configuration. This last configuration is presented in Figure 11 with the mechanical device designed to manually register the hydrofoil submergence and angle of attack during the towing tank tests. The design of this alternative configuration provided sufficient dynamic stability according the updated CFD calculations and this without changing the area or the load distribution of the V-hydrofoil configuration that is developing the same lift of the twin anhedral configuration at the same submergence. The reason of the dynamic stability enhancement is to be attributed to the different coupling between the free surface wake and the, particularly for a change of hydrofoil submersion due to a change in trim angle. In fact, in the anhedral configuration (Figure 9) due to the quasi-vertical tangent to the free surface where hydrofoil pierces the free surface, the submerged area of the hydrofoil is only very-slowly changing with a change in boat trim angle. A much more consistent change of submerged area, instead, will happen for the centered V-configuration (Figure 11) in which the hydrofoil is partially submerged under the rooster tail crest of the free surface wake at centerline (Figure 11).

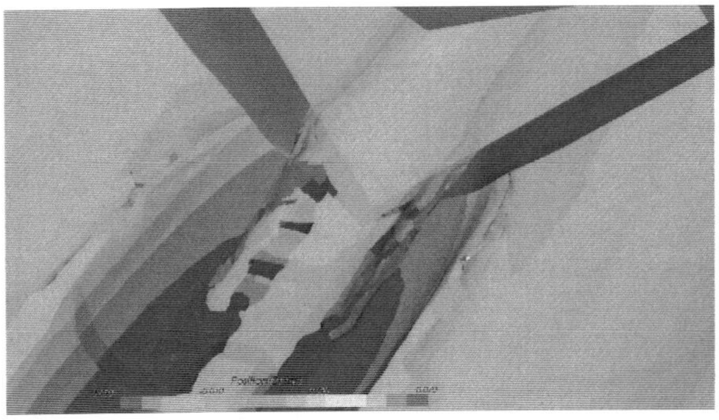

Figure 11: CFD simulation showing the final SP-SC V-shaped Hydrofoil operating in the free surface wake coming off the swept back step of the cambered planing surface.

Once again, this findings stress the importance of having some kind of method to predict the shape of the free surface in the wake of a generic highly-swept-back cambered dihedral planing surface. Current research plans are to work in this direction, with a systematic series of experimental tests at the USNA hydromechanics lab where the wake shape of different planing surface geometries will be measured in 3D. These measurements will be complemented by a series of CFD calculations, considering a more comprehensive variation of geometrical characteristics. The results will be arranged in a database to build adequate regression formulae for the free surface shape as a function of the geometric characteristics of the cambered planing surfaces. This will enable a faster and more efficient design procedure both for the aft-body and the stern hydrofoil stabilizers.

4. VALIDATION BY MODEL TESTS

A scaled down version (1:2.17) of the original 5631 model incorporating the SCPH^2 technology was design and built by MIT-iShip lab and adapted and instrumented by the Hydromechanics lab of the US Naval Academy. Hence SCPH2 model length between perpendicular Lbp=1.4m and the displacement Δ~16.5 kg.

The model was CNC machined inside (to keep the weight down) and outside from a 3D model supplied by the MIT-iShip lab and built in pitch pine wood (Figure 12, left picture). The hull was machine-finished with a fine tool, coated with epoxy resin on the inside and with a double layer of paint primer on the outside, according the standard model preparation of the Hydromechanics laboratory of the US Naval Academy (Beaver, 2014).

Figure 12: Construction of the wooden SCPH2 version of the USCG-5631model with interchangeable cambered planing surfaces (one represented in the right picture).

A horizontal flat was CNC machined on the inside at the CG location to receive the aluminum pad needed to attach the model to the towing post through the gimbal and load cell.

Figure 13: SCPH2 model suspended from the fast carriage towing post in the USNA large tank. The grasshopper used to restrain sway and yaw is visible at the bow. The V-hydrofoil stabilizer attached to the model transom through the regulating mechanism is also visible at stern.

Outside, in the bottom of the model (Figure 12, right) a recess was CNC cut in order to mount two different cambered swept-back planing surfaces. The first (A) has a standard camber distribution follows spanwise chord distribution of the cambered planing surface. The second (B) has the optimized camber distribution, found using the lifting surface code PLAMIT through a trial and error procedure, as described in the design section of the paper.

The model was towed in the large basin (116 x 7.9 x 4.9 m) of the Hydromechanic laboratory of the US Naval academy, by the fast carriage at speeds ranging from 5.5 to 9.5 m/s, corresponding to volumetric Froude numbers from 3.5 and 6.0, for the SCPH^2. To avoid a peculiar bow suction phenomena happening at the beginning of the runs, probably induced by the high impulsive starting acceleration of the carriage, the model was initially suspended to reach at a small draft in water and then released immediately at the beginning of the steady velocity part of the run. The measured drag, trim and sinkage have been averaged on the steady part of the signal identified for each individual run.

A single displacement was tested (16.5kg, corresponding to 375lbs of the original 5631 hull that was 2.17 time longer), at three different LCG position: 40%, 42% and 44% of L_{BP}, from the transom.

For each LCG position, preliminary series of runs were conducted in order to find the optimal angle of attack and submergence of the SC-SP hydrofoil stabilizer. These two hydrofoil parameters were systematically adjusted until the towed model assumed the design running trim angle τ=3.5 deg at the design speed ($Fn\nabla$=5.5). In general, this condition also corresponded to the minimum measured value of the total resistance.

An identification code was created to distinguish between the various settings: f-y+x.x, whre y is the finite draft increment given to the hydrofoil (each finite increment measures 0.5mm and is positive for a upward move) with respect to the design position (0). x.x, instead, is angular rotation of the stern SCSP hydrofoil (positive for larger a.o.a,) in degrees from the design value (0).

Figure 14: SCPH2 model towed at mazimum speed (Fn_∇=6.0) in the large basin of the USNA. LCG=42%.

Of the many runs performed during the tests, we concentrate here on the results obtained for the SCPH with camber A and LCG=42%. This condition, in fact, is directly comparable with the resistance measured on the original 5631 hull, also tested at the same LCG. The hydrofoil setting optimization came up with f-2+2.5 and this setting was fixed and used for all the other towing speeds. Indeed, the stern stabilizer could be adjusted to improve the performance at each speed and in particular at lower speeds, but we decided to investigate the performance of the $SCPH^2$ concept with a passive stern stabilizer. The picture in the front matter of the paper shows the SCPH model towed at the design speed Fn_∇=5.5. Figure 14, instead, shows the model attitude and free surface wake at the top speed Fn_∇=6.0. In both cases, the model shows a moderate trim angle (very close to the 3.5 deg design value), a fully ventilated afterbody, and a characteristic spray pattern of planing hulls well described in Savitsky and Morabito (2011) and reported in .

For the SCPH2 both the wisker spray and the spray blister seem to include a higher flow rate than in conventional planing hull and for sure a higher transversal speed component, induced be the camber. The crest of the bow "wave" is also generally higher, as well visible in Figure 14 due to the higher overpressure induced by the camber distribution on the planing surface. The same picture well evidences the shape of the free surface wake interesting the stern hydrofoil stabilizer which presents two quasi-vertical walls on the sides and a rooster tail crest at centerline.

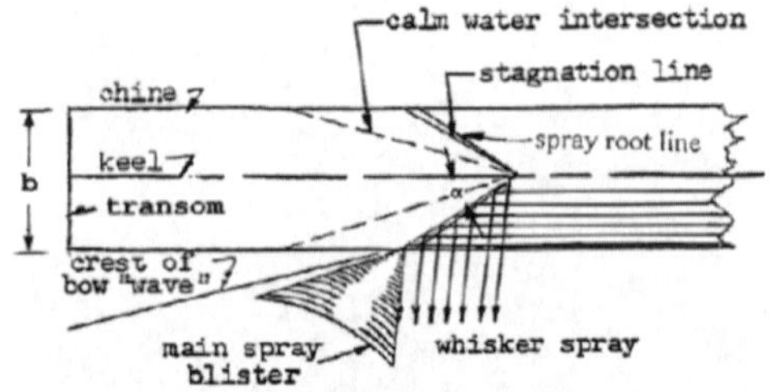

Figure 15: Spray pattern characterization of a planing hull, after Savitsky and Morabito (2011).

This characteristic flow pattern is very well captured by the underwater photo reported in Figure 17 that is relative to the design condition. The stagnation root line matches quite well the leading edge of the cambered portion of the planing surface. This is an indication of the correctness of the theoretical procedure used for the design of the $SCPH^2$ variant of the hull. The double proof is comes from the comparison of the CFD prediction with the flow pattern observed in the experiment. The matching is extremely good. Not only the position of the stagnation root line coincides well betweent he numerical and experimental tests (both run with free dynamic attitude), but also the smaller details coincide, such as the spray blister originating at midspan of the cambered bottom.

The underwater photo of Figure 17 also evidences the full ventilation of the afterbody: the free surface detaching from the edge of the step can be distinguished from the other flow regions because it presents a kind of wrinkles-like pattern.

The resistance measured during this run series is reported as a function of the advance speed in Figure 16, where the resistance of the original hull 5631, scaled down to the same size of the $SCPH^2$ model has been included for comparison.

As evident from the comparison the SCPH2 shows dramatic resistance reduction in the high speed range, i.e. Fn_∇>4.5, when the after-body starts to be ventilated.

The full ventilation of the after-body happening at higher speeds and the production of lift by camber and not by angle of attack is responsible of the drastic change of slope in the resistance curve above Fn_∇>4.0. This concept is well capture by the comparison of the calculated flow pattern for the original hull 5631 in

Figure 18 and the SCPH in Figure 17. Not only the wetted portion of the bottom has substantially reduced, but also the maximum pressure coefficient has increased due to camber and the high pressure is more evenly distributed all across the pressure area. The pressure distribution on the original hull 5631, instead, shows the typical peak at the stagnation line that rapidly drops to atmospheric pressure in the rest of the wetted bottom.

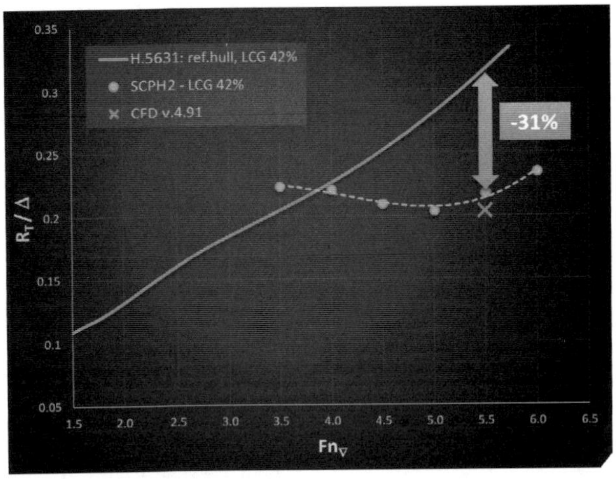

Figure 16: Comparison of model scale total resistance over displacement for the $SCPH^2$ (camber A) and the original model 5631, same displacement (36.34lbs), same LCG=42%. CFD predicted Resistance is also included at design speed.

Very remarkably:

the resistance reduction achieved by the $SCPH^2$ design variant when compared to the original 5631 hull, at the design speed Fn_∇=5.5, amounts to more than 30%

This is indeed a large number in the world of naval architects.

5. CONCLUSIONS

Without repeating design considerations and principles already given in the design section of the paper, here we would like to remark only few significant conclusions originated from the numerical and experimental studies.

High aspect ratio, cambered, stepped planing surfaces are remarkably more efficient than current prismatic (or quasi-prismatic) low aspect ratio deep-V planing crafts.
An efficient design of a Stepped Cambered Planing Hull with Hydrofoil ($SCPH^2$) variant of a modern deep-V (β=25deg) planing craft has been developed and described in the paper. Its hydrodynamic performance and advantages over the reference hull have been proven by model tests. This is the first time, at the authors knowledge, that the comparison between a SCPH2 technology is consistently compared with the original hull form.

Comparison of model tests results show that the SCPH2 technology is able to reduce the model scale drag of the reference hull by more than 30%. This is indeed a remarkable drag reduction. The reduction in full scale size will be somehow less pronounced (depending on the actual size of the vessel) but it will maintain the two digit figures.

Future research will concentrate and on the development of theoretical and numerical method to improve the design of the cambered planing surface and of the stern hydrofoil stabilizers, with the aid of high quality experimental tests. A key in this respect is the development of a low order method for characterization of the free surface in the wake of a generic cambered dihedral planing surface and new "wake-adapted" designs of SC-SP stern hydrofoils able to extend the efficiency range, ensuring an adequate dynamic stability to the whole system at the same time.

MIT-iShip lab has also completed and tested a second design of $SCPH^2$ technology, based on a different planing hull form considered as the current state of the art in ultra-fast attack crafts. The $SCPH^2$ design proved its superior performance also in this case over an even larger range of operating speeds.

6. AKNOWLEDGEMENTS

This research work has been possible thanks to the support of the US Navy Office of Naval Research, grant N00014-13-1-0332 (MIT, P.I. Stefano Brizzolara) and grant N00014-17-1-2344 (VT, P.I. Stefano Brizzolara). The SCPH2 hull concept is patent pending.

7. REFERENCES

1. Acampora, B. (1995). SM Racer: Design and Operation of One of the World's Fastest Monohulls. Marine Technology, Vol. 32, No. 3, July 1995, pp. 197-208

2. Beaver W. (2014). Details to Consider when Constructing Planing Boat Models. USNA Hydromechanic lab. Memorandum, May 2014.

3. Brizzolara S., Judge C., Beaver W. (2016). High Deadrise Stepped Cambered Planing Hulls with Hydrofoils: SCPH2. A Proof of Concept. Fifth Chesapeake Power Boat Symposium, Annapolis (MD), SNAME.

4. Brizzolara S. Vernengo G. (2016). A Three-Dimensional Vortex Method for the Hydrodynamic Solution of Planing Cambered Dihedral Surfaces. Engineering Analysis with Boundary Elements, Vol. 63, Feb. 2016, pp 15-29.

5. Brizzolara S. (2015). A New Family of Dual-Mode Super-Cavitating Hydrofoils. Proceedings

of Fourth International Symposium on Marine Propulsors, smp'15, Austin, Texas, USA, June 2015.

6. Brizzolara S., Bonfiglio L. (2015). Comparative CFD Investigation of the Performance of a New Family of Super-Cavitating Hydrofoils. Journal of Physics: Conference Series (JPCS), Vol. 656:1, pp. 12147-12150.

7. Brizzolara, S., Federici, A. (2013). Designing of V-Stepped Planing Hulls: CFD in Support of Traditional Semi-Empirical Methods". Design and Construction of Super & Mega Yachts, Royal Inst. of Naval Architects, 8-9 May 2013, Genoa, Italy.

8. Brizzolara S., Federici A (2010). CFD Modeling of Planning Hulls with Partially Ventilated Bottom. The William Froude Conference: Advances in Theoretical and Applied Hydrodynamics – Past and Future. Portsmouth, 24-25 Nov. 2010, Royal Institution of Naval Architects, vol. 1, ISBN/ISSN: 978-1-905040-77-3.

9. Celano, Tullio III (1998). The Prediction of Porpoising Inception for Modern Planing Craft. Trident Scholar Project Report, Annapolis, MD: U.S. Naval Academy, 1998.

10. Clement, E.P. (2005). A Configuration for a Stepped Planing Boat Having Minimum Drag (Dynaplane), II Ed. Private Communication; also available on the website of the International Hydrofoil Society (www.foils.org).

11. Clement, E.P., and Koelbel, J.G., Jr. (1993). Progress During the Past Century Toward the Development of Efficient, Load-Carrying, Stepped Planing Boats, Fifth Biennial Power Boat Symposium, SE SNAME Section, Feb.

12. Clement E.P. (1964). A Lifting Surface Approach to Planing Boat Design, DTMB Report 1902, Sept. 1964.

13. Clement E.P., Pope J.D. (1961). Stepless and Stepped Planing Hulls Graphs for Performance Prediction and Design, DTMB Report 1490, Jan. 1961.

14. Faison L. (2014). Design of a High Speed Planing Hull with a Cambered Step and Surface Piercing Hydrofoils. Master Thesis in Naval Engineering. MIT Innovative Ship Design lab, Dept. of Mechanical Eng., MIT, Cambridge, MA.

15. Metcalf B., Faul L., Bumiller E., Slutsky J. (2005). Resistance Tests of a Systematic Series of U.S. Coast Guard Planing Hulls. Report NSWCCD-50-TR-2005/063, Dec. 2005.

16. Moore W.L. (1967). Cambered Planing Surfaces for Stepped Hulls Some Theoretical and Experimental Results. DTMB Report 2387, Feb 1967.

17. Savitsky D., Morabito, M.G. (2010). Surface Wave Contours Associated with the Forebody Wake of Stepped Planing Hulls. Marine Technology, SNAME, Volume 47, Number 1, January 2010, pp1-16.

18. Savitsky D., Morabito M. (2011). Origin and characteristics of the spray patterns generated by planing hulls. Journal of Ship Production and Design, 27(2):63–83, 2011.

19. Savitsky, D. (1964). Hydrodynamic Design of Planing Hulls. Marine Technology, SNAME, Vol.1:1, pp 71-95.

8. AUTHORS BIOGRAPHY

Stefano Brizzolara is associate professor in the Kevin T. Crofton Department of Aerospace and Ocean Engineering at Virginia Tech and research scientist at MIT. Master in Naval Architecture and Marine Engineering from the University of Genova and PhD in Numerical Hydrodynamic for Ship Design (2000). He developed a solid ship design background in Fincantieri, where had the privilege to design cutting edge navy vessels, and a series of MDV class of fast ferries, the largest monohulls ships sailing at 40+ knots ever built in the world, predecessors of the US Navy Littoral Combat Ships. His academic career starts in the Department of Naval Architecture of the University of Genova where, in 2003, he created the new course in Numerical Hydrodynamics for Ship Design and founded the Marine CFD Group, a team of naval architects devoted to the development, validation and application of different Computational Fluid Dynamics Methods to ship design. In 2011 he joined the Massachusetts Institute of Technology as Peabody Visiting Associate Professor (MechE) and since 2013 he has been Assistant Director for Research at the MIT Sea Grant, where he founded the Innovative Ship Design Lab, iShip. He and his research group have recently moved at Virginia Tech, where they continue to focus on the development of innovative design and design tools of ocean vehicles and systems.

The hydrodynamic designs developed within the group cover unconventional high performance hull forms, like high speed SWATH, semi-displacement multi-hull vessels, stepped cambered planing hulls, super-cavitating surface-piercing hydrofoils, high efficiency low-noise propulsors and wave energy converters. Principal research sponsors are the Office of Naval Research (ONR) of the US Navy, the Defence Advanced Research Projects Agency (DARPA) and the shipbuilding industry. Dr. Brizzolara is author of more than 200 scientific papers, inventor of six patents (one pending) of marine vehicles/systems concepts.

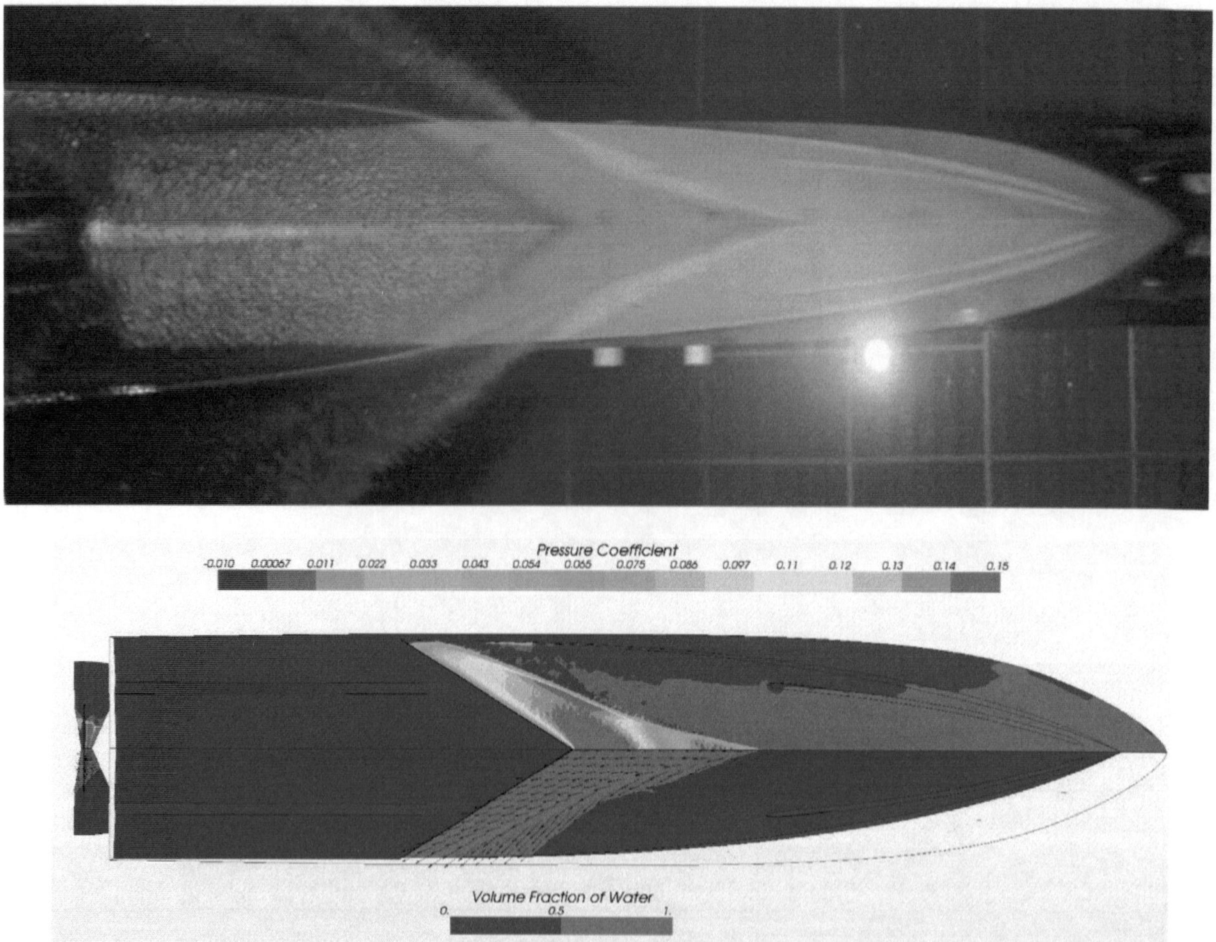

Figure 17: Top half: Underwater photo of the SCPH2 model, with cambered planing surface A, foil setting f-2+2.5, Fn_V=5.5; Bottom half: CFD simulations (same view and test condition). Upper half pressure distribution, bottom half volume fraction of water and flow direction vectors calculated in the wetted area of the hull and the stern SC-SP hydrofoil. Wetted area includes the pressure and spray area, separated by the spray root line that can be easily identified from the analysis of the calculated flow vectors.

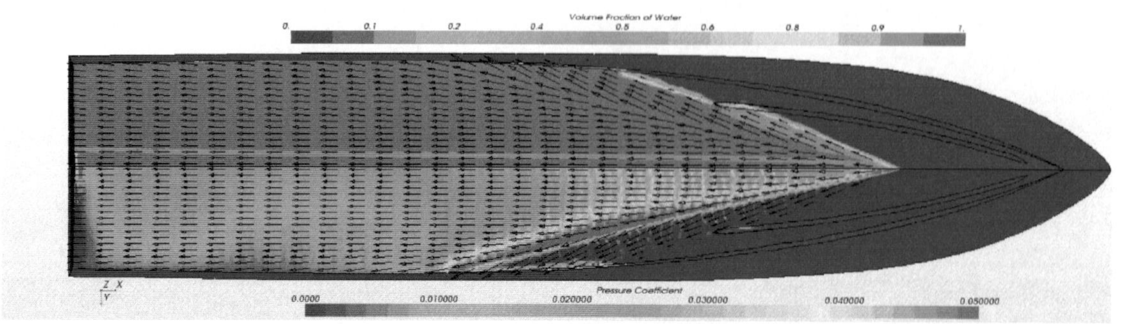

Figure 18: USCG model 5631. Results of the CFD simulations, same speed and loading condition of the SCPH2 case in Figure 18. Top half: volume fraction of water. Bottom half: distribution of pressure coefficient. The large difference of wetted area, maximum pressure coefficient magnitude and distribution between this case and the equivalent SCPH2 design (Figure 18) should be noted.

SEEKING EFFICIENCY IMPROVEMENTS FOR A MOTOR-YACHT TRANSITIONAL HULL FORM

R G Shaw, Unitec Institute of Technology, New Zealand

SUMMARY

Experience in developing and refining custom high-performance race yachts unrestricted by rules has created an innovative approach to evaluating design challenges and considering potential solutions. This methodology has been applied to developing an efficient motor-yacht hull form with the potential for increased efficiency when operating in the transitional zone between displacement and planing speeds.

This paper looks at the challenges to efficiency posed by increasing performance of a 50 metre super yacht beyond displacement speed and into the high-resistance transitional speed range. A review of existing solutions to this problem are considered, including various hull forms, use of foils and developments in multihull design. It also proposes a concept hull design, influenced conceptually by some fast displacement multihulls, which offers a solution to improved efficiency in the transitional zone.

The concept hull is developed and compared to more conventional monohull displacement and planing hull forms. Particular attention is focused on the performance in the transitional speed zone. Factors influencing performance such as appendage drag, displacement-length ratio and power-to-weight ratio are considered. Further consideration is also given to other characteristics influenced by this design concept such as safety, comfort, maintenance and directional and static stability.

1. INTRODUCTION

There is a constant drive for improvement through innovation across all forms of technology. This is true also for super yachts, where continual design refinement is sought.

The focus of this paper is on a hull form that can offer improvements in efficiency in the transitional speed range between displacement mode and fast planing. Much attention and refinement has gone into optimising both displacement hulls and fast planning hull forms. However, the difficulty of designing or developing improvements to hulls which will operate in the transitional speed zone between displacement and fast planing has seen the optimisation of hulls operating in this zone to be somewhat neglected.

With advancements in technology both providing the ability to build lighter, stronger structures and enabling lighter and more compact power plants, a great opportunity exists to reconsider how we, as designers, can create a more efficient hull that takes advantage of these developments. Another advantage of a lower displacement is the reduced regulatory framework for super yachts with displacement under 500 gross tons.

This paper considers the weight reductions that could be achieved, and how those savings could translate to improved performance. The outcome is a higher quality, higher technology and higher specification motor-yacht that has improved efficiency, reduced running costs and is potentially more sustainable and environmentally friendly.

2. BACKGROUND

Recreational powerboat design owes much of its progress to the development of the diesel engine. Initially, recreational boats were limited to displacement speed operation due to horse-power, cost and weight limitations.

Following the Second World War, with the advancement in high-power engines, boat builders and designers were quick to adapt the new technology to recreational boats. The 1960s was a golden era of advancement for performance powerboats, during which time much of the development in fast planing hull forms that we benefit from today took place. From this time there has also been continued development in displacement hull forms, seeking to refine the concept to provide the best possible sea-keeping and efficiency characteristics.

This development hasn't been isolated to monohull design; it has also produced some innovative solutions in multihulls, both power and sail. However, in spite of all of these developments there has not been a lot of advancement in motor yachts over 22 metres LOA. While some exceptions do exist, these tend to be outliers. The reality is this class of yacht is underrepresented in terms of achieving sustainable, efficient performance gains.

This paper looks at a potential hull form which, through the blending of advanced composite construction with modern power plants and innovative design, has the potential to advance the performance of this type of yacht.

3. DESIGN CRITERIA AND OBJECTIVES

For this new design it was important to frame its priority objectives. This is not intended to be a radical design that signals a major departure from current super-yacht form or function. The purpose is to explore an alternative approach to developing an efficient form that would also have positive consequences for the function of the motor-yacht.

Essentially, the main design criteria was for a 50 metre motor-yacht of moderate proportions, powered by a single screw and with the potential to efficiently operate at the upper end of the displacement speed range, approaching a Froude number of 0.6. A critical component of this was to target a lighter displacement than conventional motor-yachts of this size, through the use of composite construction throughout.

The key objective for this design is for the motor-yacht to have the potential to efficiently operate for extended periods at speeds considered higher than those associated with conventional displacement mode operation.

4. REVIEW OF CURRENT SOLUTIONS

Considering the development of a modern performance monohull which could exceed the traditional constraints of displacement operation, a range of current solutions were identified and considered. Strikingly, all of the options identified tend to fall in to either the conservative or extreme categories.

For displacement hulls, there are a few basic variations on the same theme, which are clearly very functional. At the other end of the range, where high speeds are the priority, this characteristic is often pursued at great expense. This expense is multifaceted and can be measured in financial cost (both build and operational), complexity and often also practicality and some would say aesthetically. Despite these developments, there is an apparent neglect of developing a hull form which combines the technologies currently available to achieve the performance advancements desired.

The use of foils on performance yachts is gaining attention, with excellent results achieved in dinghies such as the International Moth, ballasted monohulls such as the Quant 23 and IMOCA Open 60 class, and now the America's Cup multihulls. Applying such foils to a motor-yacht is possible, although complex, and their performance is very speed-dependent.

For a vessel operating in the speed range being considered in this paper, it is considered that conventional lifting foils would not offer a significant performance enhancement. Certainly, when considered together with the added expense and complexity of such systems, it is not an attractive option for a vessel of this type.

Alternatively, the use of a single foil such as the Hull Vane® developed by Van Oossanen Naval Architects is a more practical solution for a vessel operating in this targeted speed range. This patented foil is positioned below the waterline, near the aft-body of a vessel, and is aimed at regaining part of the energy lost in the transom wave. It also influences the trim, and is designed to reduce the vertical motions of the vessel. [1]

This clever and effective appendage is best suited to achieve performance gains in the Froude number range of 0.2 to 0.7, which is similar to the speed range targeted by this design. The Hull Vane® shows great potential and has the advantage of the potential to be retrofitted to existing boats. However, the solution investigated in this paper also has the potential to achieve many of these same benefits without the complexity and exposure of adding a wing to the aft body of a vessel.

5. DESIGN ADVANTAGES

Principally, this motor-yacht design explores the concept of a canoe-stern underbody appendage which partially accommodates the engine and drive train. The purpose of this feature is to enable the engine and drive train to be positioned lower than the hull fair body. Through lowering the engine and drive train below the hull there are several potential advantages that can be achieved over a conventional hull and drive train geometry. It allows the potential to design a hull form that isn't compromised by the need to accommodate the engine, drive train and appendages. This allows more freedom in the design of the hull, to better achieve an efficient hull form for the intended operating speed range.

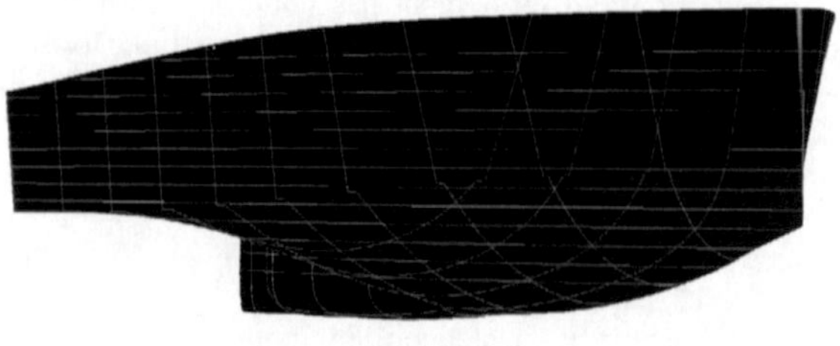

Figure 1: Candidate hull model, viewed from the starboard forward quarter.

By lowering the engine and drive train down into the underbody appendage it is possible to align the shaft angle with the direction of water flow, which gives a reduction in drag due to eliminating the exposed inclined propeller shaft and eliminating the need for a shaft support strut. Further performance improvements are gained through aligning the propeller with the water flow, improving the performance and efficiency of the propeller. With developments of the appendage geometry it could also be possible for the appendage to provide protection for the propeller and rudder in the case of grounding or a collision with a submerged object.

The engine appendage can also contribute to the directional stability of the vessel in the same way as a conventional deadwood or keel. This can be beneficial in quartering or crossing seas.

Through lowering the engine down into the appendage the vertical centre of gravity of the vessel is also lowered, improving the vessel's range of stability. A secondary potential advantage is to lower the accommodation space within the vessel to achieve a more comfortable motion at sea and improve the connection with the environment at rest.

6. PERFORMANCE

Several attributes are considered and refined in an effort to increase the performance profile of the motor-yacht. A single-screw propulsion configuration is combined with the propulsion appendage arrangement and minimising the structural weight through the use of carbon composite construction.

Utilising single-screw propulsion is a significant aspect of the weigh-saving concept of this yacht. In optimising the vessel to perform at the relatively low speed range of up to Froude number 0.6, the major contributors to resistance need to be considered and prioritised.

It is well understood that at this speed range, beam-to-length ratio and displacement have a significant impact on resistance, as do frictional drag, form drag, induced drag and aerodynamic drag. To reduce resistance and, probably most importantly, minimise wave drag, this design focuses on reducing displacement, a moderate beam-length ratio and a narrow angle of entry.

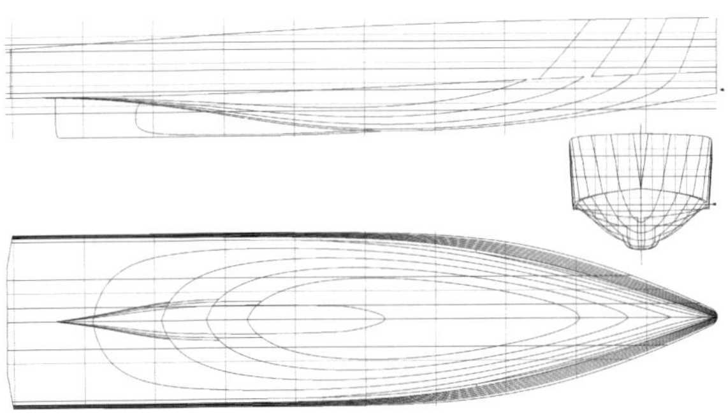

Figure 2: Candidate hull model lines plan.

Above Froude number 0.3 to approximately 1.2 is where the vessel enters the transitional speed zone, and it is here that wave drag is the most significant component of resistance. This hull form enables a warped plane hull with a fine entry and flat aft sections, to prevent squatting and maximise pressure recovery. It is possible to have the performance benefits of this type of hull but also benefit from reduced appendage drag and improved propeller efficiency due to the engine appendage.

It is accepted that there is some disadvantage due to the increased wetted surface area of the engine appendage. How-ever, given that wave drag is the major drag contributor at this critical speed range, it is a small price to pay in light of the advantages it offers.

Hydrostatics

Measurement	Value	Unit
Displacement	384.8	t
Volume (displaced)	375.389	m^3
Draft Amidships	2.29	m
Immersed depth	2.61	m
WL Length	49.982	m
Beam max extents on WL	9.581	m
Wetted Area	489.799	m^2
Max sect. area	13.436	m^2
Waterpl. Area	356.202	m^2
Prismatic coeff. (Cp)	0.559	
Block coeff. (Cb)	0.3	
Max Sect. area coeff. (Cm)	0.603	
Waterpl. area coeff. (Cwp)	0.744	
LCB length	-27.231	
LCF length	-30.321	
LCB %	-54.482	
LCF %	-60.663	
Immersion (TPc)	3.651	tonne/cm
MTc	10.755	tonne.m
RM at 1deg = GMt.Disp.sin(1)	34.346	tonne.m
Length:Beam ratio	5.217	
Beam:Draft ratio	3.67	
Length:Vol^0.333 ratio	6.929	

Figure 3: Candidate hull hydrostatics.

7. APPENDAGE DESIGN

A range of appendage design variations were considered. Inspiration was drawn from the Malcom Tennant canoe stern (CS) multihull design concept. [2] To achieve the positive benefit of the Tennant CS design, a different approach was necessary to adapt the concept to a monohull hull form.

The appendage solution applied in this design is an adaption of a proprietary low-drag ballast bulb section developed for racing yachts. This section has been modified to provide suitable proportions to contain the engine and drive train while morphing from the hull form, giving a fair transition. The aft sections of the appendage have also been modified to provide a profile that allows the lowest drag shaft exit and potential for the addition of a rudder skeg support.

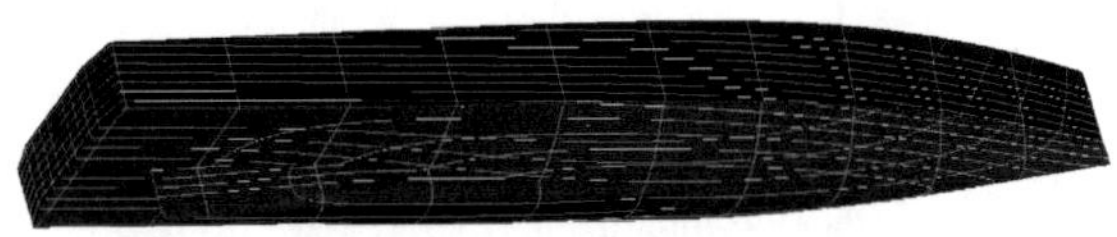

Figure 4: Candidate hull starboard rotation showing bilge appendage.

8. CONSTRUCTION

A critical component of the success of this hull form is the weight reduction achievable through the utilisation of building in carbon composites.

Advanced composites have been in use in the aerospace industry since the 1960s. Since then, the use of this construction method has become more widespread, and it has been undertaken in the marine industry for several decades now. In the 1980s, the use of advanced composites was adapted for boatbuilding in the production of IOR race yachts. It came to prominence in the construction of the New Zealand America's Cup challenger *KZ7*, which was runner-up in the challenger series in 1987.

Since then, advanced composite construction has been fully utilised and adapted to marine construction, building boats from the single-person foiling Moth to America's Cup yachts, both mono and multihull, and super-yachts. *Hetairos III*, at 66.4 metres LOA, is the largest super yacht so far built in carbon composite.

Improvements in processing techniques such as vacuum bagging, infusion and the use of pre-preg laminates has made composite construction more attractive and practical. These improved techniques have increased the quality and reliability that builders have been able to achieve.

Since it was first introduced to the marine industry, the cost of carbon composite laminate has dramatically reduced, making it a much more viable option for the construction of large structures. These improvements in cost, together with the weight savings, increased durability and longevity they offer, make composites a viable build medium for larger motor-yachts.

The use of composites eliminates the limitations on producing complex geometric shapes. This can be an advantage when designing both small and large surfaces, giving the designer unrestricted potential to explore more practical or aesthetically pleasing geometries.

While reducing the total weight of the yacht has obvious performance advantages, there are also other advantages in creating lighter components. The ability to reduce the weight of various hatches, doors and other movable components is significant. These advantages can range from reducing the need for electric/hydraulic operating mechanisms as well as reducing the potential risk of personal injury through operating these movable components.

Well-engineered and -manufactured composite structures are very long-lasting and are less susceptible to many of the corrosion risks prevalent in the marine environment. Composite structures have low fatigue rates and good resistance to chemical and electrolytic corrosion.

Traditionally, the complexity associated with building one-off composite vessels has made it not economically viable to pursue. The need to build a timber plug and finish it to a suitable standard to laminate the structure over was labour-intensive and added to the build time and workforce skills requirements. More recently, through the advancements in CAD-CAM technologies, larger CNC machining capabilities and the expanding capabilities of composite boatbuilders, it is now a viable option to build one-off composite boats. [3]

For this motor-yacht it is envisaged the structural surfaces would be built from carbon laminates and utilise a foam core. For the engine appendage, the use of monolithic construction is anticipated; the shape is well suited to this method but also it offers the most robust and resilient structure. There will be significant loads acting on this structure and it will need to be very stiff to ensure no vibration is caused through the propeller and drive train. It is also recognised that in the case of a grounding or collision with a submerged object, it is this part of the vessel that is likely to be at highest risk.

9. ANALYSIS

At the current concept stage, analysis has been conducted using three prediction analysis algorithms: the Savitsky pre-planing, Savitsky planing and the Holtrop methods.

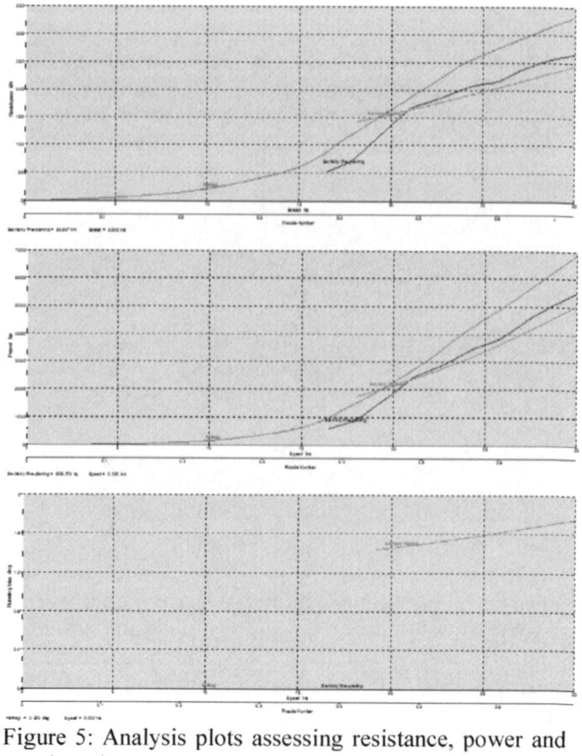

Figure 5: Analysis plots assessing resistance, power and running trim.

While applying these algorithms has provided useful and valuable analysis of the potential performance of this hull form, the application of these algorithms is considered solely a first step in analysing the performance of this concept. The next steps are to undertake extensive CFD analysis using the openFOAM software package. A focus on hull form, appendage design and the hull/appendage interface will be explored. Attention to ensuring an undisturbed flow of water onto the propeller will be a priority throughout.

Following a successful CFD study, development and testing of a scale model of the concept will be pursued.

Speed kts	Froude No. LWL	Froude No. Vol	Savitsky Pre Planing Resist (kN)	Savitsky Pre Planing Power (kW)	Savitsky Planing Resist (kN)
0	0	0	--	--	--
0.75	0.017	0.046	--	--	--
1.5	0.035	0.092	--	--	--
2.25	0.052	0.138	--	--	--
3	0.07	0.183	--	--	--
3.75	0.087	0.229	--	--	--
4.5	0.105	0.275	--	--	--
5.25	0.122	0.321	--	--	--
6	0.139	0.367	--	--	--
6.75	0.157	0.413	--	--	--
7.5	0.174	0.459	--	--	--
8.25	0.192	0.504	--	--	--
9	0.209	0.55	--	--	--
9.75	0.227	0.596	--	--	--
10.5	0.244	0.642	--	--	--
11.25	0.261	0.688	--	--	--
12	0.279	0.734	--	--	--
12.75	0.296	0.78	--	--	--
13.5	0.314	0.825	--	--	--
14.25	0.331	0.871	--	--	--
15	0.349	0.917	--	--	--
15.75	0.366	0.963	--	--	--
16.5	0.383	1.009	61.2	696.67	--
17.25	0.401	1.055	73.44	873.98	--
18	0.418	1.1	85.84	1065.98	--
18.75	0.436	1.146	112.24	1451.85	--
19.5	0.453	1.192	138.65	1865.24	167.43
20.25	0.471	1.238	164.09	2292.33	173.6
21	0.488	1.284	189.34	2743.01	179.93
21.75	0.505	1.33	203.89	3059.31	186.4
22.5	0.523	1.376	212.65	3300.86	193.03
23.25	0.54	1.421	222.58	3570.04	199.79
24	0.558	1.467	233.81	3871.25	206.68
24.75	0.575	1.513	243.1	4150.89	213.7
25.5	0.593	1.559	247.54	4354.79	220.85
26.25	0.61	1.605	252.81	4578.32	228.11
27	0.627	1.651	264.96	4935.44	235.49
27.75	0.645	1.697	277.13	5305.44	242.96
28.5	0.662	1.742	285.29	5609.18	250.54
29.25	0.68	1.788	293.13	5915.12	258.21
30	0.697	1.834	299.48	6198.17	265.96

Speed kts	Froude No. LWL	Froude No. Vol	Savitsky Planing Power (kW)	Holtrop Resist (kN)	Holtrop Power (kW)
0	0	0	--	--	--
0.75	0.017	0.046	--	0.15	0.08
1.5	0.035	0.092	--	0.56	0.58
2.25	0.052	0.138	--	1.2	1.86
3	0.07	0.183	--	2.05	4.24
3.75	0.087	0.229	--	3.12	8.07
4.5	0.105	0.275	--	4.4	13.65
5.25	0.122	0.321	--	5.88	21.3
6	0.139	0.367	--	7.58	31.36
6.75	0.157	0.413	--	9.5	44.22
7.5	0.174	0.459	--	11.67	60.4
8.25	0.192	0.504	--	14.17	80.63
9	0.209	0.55	--	17.05	105.88
9.75	0.227	0.596	--	20.42	137.34
10.5	0.244	0.642	--	24.41	176.82
11.25	0.261	0.688	--	29.29	227.32
12	0.279	0.734	--	34.91	288.99
12.75	0.296	0.78	--	40.77	358.61
13.5	0.314	0.825	--	47	437.7
14.25	0.331	0.871	--	54.44	535.15
15	0.349	0.917	--	64.17	664.03
15.75	0.366	0.963	--	77.27	839.61
16.5	0.383	1.009	--	94.6	1076.83
17.25	0.401	1.055	--	116.06	1381.16
18	0.418	1.1	--	131.69	1635.25
18.75	0.436	1.146	--	147.49	1907.83
19.5	0.453	1.192	2252.41	163.48	2199.19
20.25	0.471	1.238	2425.25	179.64	2509.58
21	0.488	1.284	2606.72	195.98	2839.29
21.75	0.505	1.33	2796.99	212.5	3188.57
22.5	0.523	1.376	2996.21	229.2	3557.7
23.25	0.54	1.421	3204.52	246.07	3946.94
24	0.558	1.467	3422.05	260.69	4316.35
24.75	0.575	1.513	3648.91	271.69	4639.03
25.5	0.593	1.559	3885.2	282.11	4962.82
26.25	0.61	1.605	4130.99	292.18	5291.22
27	0.627	1.651	4386.37	302.11	5627.3
27.75	0.645	1.697	4651.37	312.03	5973.61
28.5	0.662	1.742	4926.04	322.06	6332.26
29.25	0.68	1.788	5210.39	332.27	6704.94
30	0.697	1.834	5504.42	342.72	7093.03

Figures 6a and 6b: Analysis data showing resistance, power and running trim

10. CONCLUSION

Through study and observation of conventional solutions to designing hulls that operate in the transitional speed range came a curiosity to explore alternative hull forms that might better operate in this range. What has been developed is a concept only and needs much refinement to be developed into a viable solution with real advantages.

It is clear, however, that not only can this concept have potential for real performance improvements but that it can also impact on life on board, both at sea and at rest. It has the potential to both lower the vessel's centre of gravity and bring the accommodation space lower to the water, further improving comfort and the connection with the environment for both guests and crew.

While advanced composite construction has become a standard feature in performance boats, both power and sail, it is still quite uncommon in super yachts. There is potential to achieve significant improvements in performance through the application of these materials. Carbon composite construction has many advantages where higher performance is sort and is now a more viable construction medium for large, one-off boats than ever before. Through combining this advanced material with an innovative hull form and the benefits it offers, potentially significant performance gains can be realised in a practical, modern motor yacht.

11. REFERENCES

1. K Uithof, PG van Oossanen and F Bergsma, 'The feasibility and performance of a trimaran yacht concept equipped with a Hull Vane®', *Design and Construction of Super and Mega Yachts*, 2015.

2. For more on this, see www.tennantdesign.co.nz

3. P Dassi, 'Advanced composites: a possible answer to mega yacht structural questions', *Design and Construction of Super and Mega Yachts*, 2015.

12. AUTHOR'S BIOGRAPHY

Robert Shaw is a lecturer and academic leader at Unitec Institute of Technology in Auckland, New Zealand, specialising in yacht design and leading academic development and research in his department. He is also the principal of performance yacht design studio Shaw Yacht Design, leading collaboration and consultation on custom performance-yacht design projects.

SUSTAINABLE LUXURY ICE-CLASS ARCTIC EXPLORER VESSEL FOR MILLENNIALS

S McCartan, EBDIG-IRC, NTDC, Coventry University, UK + EBDIG, IUAV, Venice, IT
A Barrett and **L Swinfield,** EBDIG-IRC, NTDC, Coventry University, UK
B Verheijden, Academy Minerva Groningen, NL
D Boote, DITEN, Genoa University, IT
L Mori, De Jorio Design International, Genoa, IT
R Schouten, Feadship, NL

SUMMARY

Luxury has undergone a paradigm shift from materialism to experiential value. Millennials crave new and unique curated experiences, to see a place that few or none have seen. Sustainable luxury is the new luxury agenda, it offers the opportunity of a new design meaning for the next generation of gigayachts through an holistic approach encompassing a range of issues from addressing pollution to societal contributions. Arctic ice shrinkage is a global concern producing an opportunity for unique luxury experiences and urgently needed scientific research. This Design-Driven Innovation project presents a 300m sustainable luxury Ice-Class Arctic explorer vessel which supports scientific research and curated luxury experiences. It combines a futuristic exterior form with the interior urban flow of a cruise ship strongly influenced by architectural trends, resulting in the spectacularisation and conquest of landscape normally associated with cruise ships. Where drones capture a unique social media narrative of a once in a lifetime experience. The business model is based on private members club such as Quintessentially One. To support the sustainability agenda the vessel and tenders are hydrogen powered. The hydrogen is produced from renewable energy through a network of floating wind turbines located throughout the region.

1. INTRODUCTION

This paper presents a conceptual design approach, as proposed by McCartan et al [1], to support the evolution of the gigayacht towards a less egocentric product. The design approach is a Transfer of Innovation (TOI)informed by the design practices of the cruise ship industry. The design proposal presented addresses the market sector between superyachts and ultra-luxury cruise ships for millennials, through a new design meaning informed by technological innovation. The design approach engages in the process of Design-Driven Innovation. Contemporary trends in cruise ship are rapidly evolving towards spectacularization and dematerialization. Entertainment is a core focus for cruise companies, immersing the passengers in a range of emotionally engaging experiences, which continuously change in space and time, resulting in the ever growing investments in entertainment technology. The passenger is attracted in an emotional path, constructed to give surprise and delight at every corner, through formal stereotypes, pop culture, and live shows. Resulting in the creation of a continuous passengers' flux along the ship in an urban-like environment, where different design languages meet.

1.1 MILLENNIALS AND LUXURY

Millennials include anyone born between the early Eighties, an era predating the use of mobile phones, and the early 2000s, the Internet age. They are a free-thinking, individualistic generation, currently representing a consumer population of over 80 million. By 2035, Millennials will have the potential to become the largest spending generation in history. Millennials' influence will be felt by 2020 as the oldest Millennials "Millennial+", will be entering their peak earning years and will have disposable income for luxury experiences. This trend will continue as more Millennials become "Millennial+".[2]

Millennials have redefined the term luxury. Luxury no longer refers exclusively to expensive bags, shoes and watches. It now includes experiences such as farm-to-table dining and "über-luxe" travel. Millennials prefer experiences to things. Fashion brands are blurring the lines between luxury items and experiences. Luxury shopping has been developed into a curated experience. Digital platforms are creating new, interactive experiences every minute. In 2017 there has been a 400% increase from the previous year in the Instagram posts of fashion brands. Designers are focused on giving Instagram influencers preferential treatment, including access to their shows and collections. In doing so they are trying to inculcate in millennials an emotional connection with their products . [2]

Millennials aren't the same as previous generations. They were raised to think very holistically, having the awareness that every decision has an impact on the world in some way. Especially, when it comes to investing and consuming. Millennials also want to be stewards of the environment; protecting it, and bringing ecosystems back to full-strength. They want society to be fair and equal. The generation wants balance in what is known as the '*Triple Bottom Line*' an accounting framework that takes into account social, environmental, and economic performances instead of simply focusing on financial performance.[3]

Previous generations were largely unaware of the severe impacts their activities were having on the environment. Decisions were short-sighted and investments often were made in environmentally harmful industries. While investing in these industries is likely to offer financial return, the non-financial impacts of these investments have been extremely harmful. Due to their environmental awareness, millennials know that industries such as oil and gas can have a significant impact on the environment. Based on this knowledge, some are less likely to invest in oppertunities that, while offering an attractive financial return, will severely harm the environment. [3]

Morgan Stanley Private Wealth Management and Campden Wealth conducted a survey [4] of 87 ultra-high net worth individuals under the age of 40, who come from families with a minimum wealth of $25 million. Over half (57%) had a family net worth of more than $100 million. Ultra-wealthy millennials are idealistic about the uses of their wealth, but say they are generally aligned with the values of their parents:

- 63% of millennials view themselves as stewards of their wealth for future generations, compared with 46% of older siblings.
- 58% view their wealth as a vehicle to help the community (vs. 38% of older inheritors)
- 74% view their wealth as a source of empowerment to pursue what is most important (vs. 54% of older generation).

The next generation members of ultra-affluent families are seeking to define their place in the family and find their voices. Education and opportunities will be central to their success. Areas such as philanthropy, values-based investing and entrepreneurism have high appeal with this group and present strong opportunities for engagement.[4]

Millennials would pay slightly more for eco-friendly products, and that logic should carry over to investments. If a yacht charter brand can show millennials that they wish to make a positive difference to the environment, they can align with their interests, wants, and needs. This method of alignment is much more effective than simply by promoting superyacht charter as high status conspicuous consumption. "Green" investment packages, avoiding oil, gas, automotive and tobacco investments, are perceived to be quite attractive to some millennials. A significant challenge in branding superyacht charter as being environmentally friendly is "greenwashing". If charter brokers claim to offer environmentally friendly vessels, or investment packages, but simply repurpose another vessel or apply incremental technological innovation and claim it to be green, millennials will notice. The group has grown up in a time of unprecedented environmental awareness, and it will damage a charter brand's reputation if they are perceived as greenwashing. [3]

A key approach to use causes to help attract millennials to superyacht charter is through content and branding. By sharing views on being inclusive, open-minded and environmentally conscious. If a charter group genuinely believe in doing right by their clients, this branding should come easily, positioning their core concerns as a part of the business. The generation cares about others. They want people to be treated the way they want to be treated and called what they want to be called. They want people to not be discriminated against based on who they are. [3]

Millennials really care about cause related initiatives. They will respond positively when a company takes interest in a social or environmental cause, with 75% of millennials believing that it is important that a company gives back to society. Millennials have surpassed simply wanting help in supporting causes and are starting to demand that others, especially companies, do their part. Millennials appreciate commitment to environmental and social responsibility. After learning that a company is socially or environmentally responsible:

- 83% are likely to trust the company more
- 79% are likely to purchase that company's products
- 74% are more likely to pay attention to that company's message because it has deep commitment to a cause [3]

An example of a company using positive social influence to sell their product is Toms shoe company. They do their part in giving back to society, by giving a child in need a free pair of shoes for every shoe sold. This is a core aspect of the company and a contributor to its rapid growth. Connecting experiences back to social and environmental responsibility place a company in a strong position to build trusting relationships with millennials.[3]

1.2 SUSTAINABLE LUXURY

In 2009 sustainability officially entered the luxury professionals' agenda. The consequences of such a dramatic acknowledgement are far-reaching and entail a revision of every single dimension of the industry, including product development, marketing, finance, distribution. This evolution of luxury will redefine the brand identities and brand perceptions. Florian Gonzalez [5] proposed the following guidelines for a more sustainable luxury industry:

- Design can be eco-driven and not ego-driven. Sustainability shall inspire new co-design and co-creative processes in the name of 'design activism' [6]

- Being carbon neutral is only a first step: actually reducing carbon emissions and making a positive impact is the key objective

- Brands which consistently and constantly improve their sustainability credentials will reinforce customer loyalty and increase their customer base over time

He proposed that sustainability requires a permanent learning process made of humility, generosity, kindness and sharing of knowledge. In which companies should not be afraid of sharing best practices and creating positive externalities. Diana Verde Nieto [7] developed the Positive Luxury website and the Blue Butterfly interactive Trustmark to be a guide for positive living by providing meaningful information to people and reward brands that invest in bettering the state of the world through consumer action. It is designed to be used at every point of the day, such as deciding what restaurant to eat at in the evening. The rationale for applying it across lifestyle is that the people that value positive fashion, also want, food, hotels and other luxury goods. This is the first interactive Trustmark that examines a brand's overall holistic appeal.

The site awards brands that fit rigorous sustainability/ethical criteria with the trustmark, to communicate what the brands do in terms of social and environmental efforts, without assigning a value judgement. It allows consumers to see for themselves and make their own choices based on style, product functionality and most significantly, ethical actions. For example, information stating that a brand employs policies to prevent the use of child labour, forced labour, encourages recycling and regularly re-evaluates its supply chain. A wide mix of around 420 lifestyle brands have registered with the site. The Bulgari hotel resort [8] engaged in sustainable luxury through implementing the principles of Passive Design, it was developed according to the best practices in sustainable design set by BREEAM [9]. A green roof has been ecologically designed to highest standards by the London Biodiversity Action plan. Further examples of green initiatives include electric car charging points to support environmentally friendly forms of transport, and 20 bicycles to guests to enjoy Hyde Park and discover London. This builds upon recent research that highlights the experiential value of luxury tourism [10]. The Sustainable Luxury Working Group (SLWG) [11] has been established by companies in the luxury industry that are committed to advancing good social, environmental, and animal welfare practices in their business operations, including sustainable sourcing practices. Their members include: Me; Cartier; Chloé; Kering; Mulberry Group; Ralph Lauren; Tiffany & Co.

The objectives of the group are: to discuss common environmental, social, and governance challenges facing the luxury sector; to promote transparency, knowledge sharing, and collaboration across common global supply chains.

1.3 ARCTIC SCIENTIFIC RESEARCH

An understanding of the causes of the ongoing rapid changes in the Arctic and their impact on the future, is a scientific challenge of the utmost urgency with important societal implications. Changes in the Arctic affect the nature and frequency of extreme weather events and other natural hazards which threaten the UK. Increasing ice-melt on the region's glaciated islands and the vast Greenland ice sheet could lead to changes in sea level which will affect communities and assets along UK coastlines. Frozen soil, known as permafrost, stores greenhouse gases which are released if it melts, further reinforcing changes brought about by global and regional warming.[12]

As the Arctic ice retreats, natural resources such as fish and mineral resources become more accessible. In addition to this the Northwest Passage may become an important new shipping route if it becomes ice-free on a regular basis. In order to effectively manage this combination of threats and opportunities, a multidisciplinary approach is required to gain insight into the scale of the future challenges for governments, indigenous communities, natural habitats and businesses keen to operate in the Arctic. The Natural Environment Research Council (NERC) Arctic Research Programme's key challenge is to understand and improve predictions of Arctic change and to establish the significance of the associated regional and global impacts. This overarching objective informs the four major interlinked scientific areas of inquiry funded by the programme:

- Large scale variations in the temperature and amounts of ice and water in the arctic along with atmospheric changes require more study and greater understanding.
- Given the potential for large-scale greenhouse gas emissions as the Arctic warms it is crucial we estimate how much carbon is stored in Arctic permafrost and continental shelves and investigate the timescales over which it might be released.
- Improving existing computer models to project what the future holds for the Arctic and the rest of the world.
- Knowledge of how stable underwater slopes in the Arctic are, how seismic activity or degassing of thawing methane hydrates within sediments could affect them and what natural hazards could be caused if they collapse.[12]

1.4 ACTIVITIES IN THE ARCTIC

Shipping in Arctic waters is predicted to increase due to climate change and the melt of sea ice. Ships usually enter the western portion of the Northwest Passage from the east through a southern route along the main coast, the Amundsen Gulf, or through two northern routes, either north of Banks Island or south-east of Banks Island. Potential environmental effects of increased shipping activities in the Northwest Passage and the Beaufort Sea include higher risk of oil or waste spills, changes in ice conditions due to leads caused by ship wakes, and impacts on wildlife and marine species.[13]

Small boats and yachts of less than 20m may have considerably less environmental impact than larger vessels. Passages made by icebreakers can use more northern routes, and can open leads in the sea. A record number of vessels, 30 in total, transited through the Northwest Passage in 2012. In 2013, for the first time, a large bulk carrier transited the Northwest Passage. Only 17 vessels managed the full northwest passages in 2014, due to a short and cold summer. Since the first crossing of the Northwest Passage by Amundsen in 1906, less than 1 ship every 10 years on average had successfully completed the full passage until 1969, when the oil tanker SS Manhattan, refitted with an ice-breaker bow, crossed the Passage from east to west, and then returned east. The number of completed trips through the Arctic Ocean increased in the late 1970s, mostly due to the availability of icebreakers and other ships capable of navigating in difficult northern waters. This is particularly the case for Arctic tourism [13]

From the 1980s on, voyages through the Passage have become an annual event. The number of transits increased from 4 per year in the 1980s to 20-30 per year in 2009-2013. These transits are mostly completed by icebreakers on coast guard and research duties, small vessels or adventurers, passenger ships offering Arctic tourism opportunities, and tug and supply vessels, some with barges, as shown in Figure 1. Other types of ships completing the passage include oil/fuel tankers, drill ships, seismic vessels, cable vessels, and buoy tenders. A great portion of the increase in transits since the late 1980s is due to an increase in shipping activities by tug-supply vessels, 50% of which had icebreaking capability, involved in the oil and gas industry in the Beaufort Sea. More recently, there has been an increase in ship-based research in the Northwest Passage and the Beaufort Sea, attributable to concern over the effects of climate change in arctic marine ecosystems, culminating in more research efforts during the International Polar Year. [13]

In August 2016 the cruise liner Crystal Serenity was the largest to sail through the Northwest Passage. Promoted as the "ultimate expedition for the true explorer," the 32-day trip departed from Anchorage, traveling north through the Bering Strait and across the Canadian Arctic, then made stops in Greenland and the northeastern United States before docking in New York City. Passenger tickets for the excursion ranged from $22,000 to $120,000, plus required insurance coverage of at least $50,000 for emergency evacuation.[14] An exponential growth of expedition cruising is expected to start in 2017. Lindblad Expeditions launches its brand-new ship, National Geographic Quest, in June 2017 in Alaska and the Pacific Northwest. Ventus Australis is a new-build from the Chilean-based expedition cruise line, Australis. With 210 passengers capacity, in 100 cabins, the ship launches in January 2018. Meanwhile, Oceanwide Expeditions' newest cruise ship, the Polar Class 6 vessel Hondius will enter service in 2019. It joins the Dutch line's existing polar expedition ships, the 116-passenger Plancius and Ortelius. Some larger cruise ships run on HFO that would be extremely difficult to clean up if spilled. International shipping is responsible for 2.6 % of all global carbon emissions, according to the IMO. The IMO expects emissions to increase by at least 50% by 2050. [15]

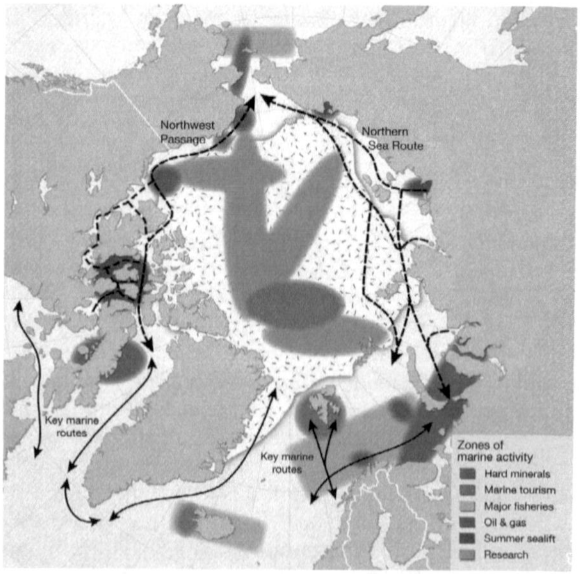

Figure 1: Zones of marine activity [13]

2. DESIGN-DRIVEN INNOVATION SCENARIO

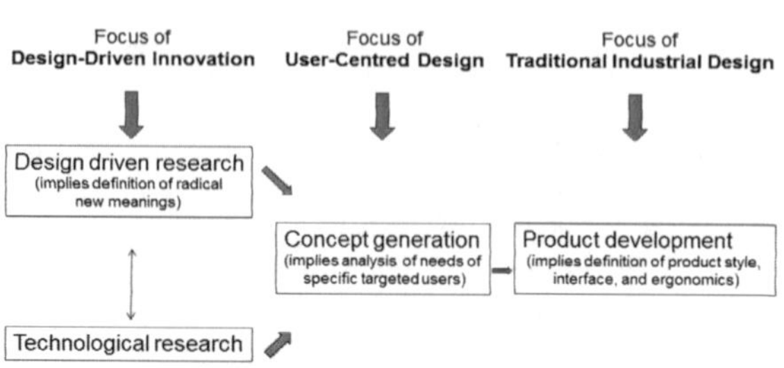

Figure 2: Design-Driven Innovation as research [17]

The perceived comfort/pleasure of a charter client or owner on a superyacht is a combination of the emotional response to the visual impact of the aesthetics of the vessel exterior/interior and that of the interaction of the user with the environment. Design must therefore meet user needs, both functional and emotional, and been informed by human factors. People gain pleasure from products that meet their functional and supra-functional needs, as defined in the most comprehensive terms. Indeed, now that adequate product functionality is the norm, supra-functional factors are being recognized as more important. Emotional bonding, symbolic representation, tribal connections, subculture references, and so on all form part of the language defining product personality and product semantics. People relate to products in individual and interesting ways. Different people relate to the same product in their own particular way, depending upon its characteristics and their own. Material possessions serve as symbolic expressions of who people are. The clothes they wear, the car they drive, the design meaning of the superyacht they own or charter all enable they to express their personality, social standing, and wealth. The nature of a product can be described as a product personality, and it is this, that determines the relationships that users develop with different products. [16]

To facilitate design innovation marine designers should consider implementing a Design-Driven Innovation strategy as is often employed within product design. People do not buy products but buy design meanings. People use things for profound emotional, psychological, and socio-cultural reasons as well as utilitarian ones. Analysts have shown that every product and service in consumer as well as industrial markets has a design meaning. Marine Designers should therefore look beyond features, functions and performance, and understand the real design meanings users give to a superyacht. The process of Design-Driven Innovation is an exploratory research project, which aims to create an entirely new market sector for a given product through changing the design meaning the user has for the product. It occurs

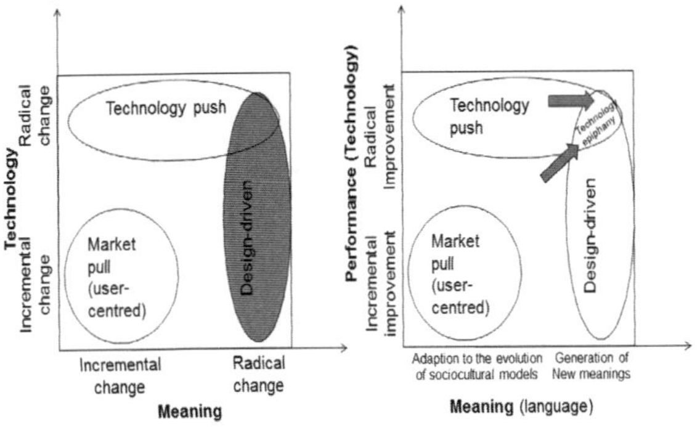

Figure 3: The strategy of design-driven innovation as a radical change of design meaning [17]

before product development, as shown in Figure 2, and is not the fast creative brainstorming sessions that are typical of concept generation but a design investigation similar to technological research. In essence, it is the development of a design scenario through engaging with a range of interpreters in technology and cultural production. Knowledge is generated from immersion with the design discourse of the interpreter's groups. The process can be structured or unstructured and is dependent upon the nature of the relationship of the client with the interpreters. The interaction between innovation of design meaning and technology innovation can transform the market within an industry and even create new market sectors. The two strategies are complimentary as technological and socio-cultural models are inextricably linked, evolving together in innovation cycles. The successful interaction between design-driven and technology-push innovation is called a technology epiphany, shown in Figure 3, it creates a market leader and in some cases a completely new market sector. It is the basis for successful products such as the Apple iPod. [17]

Luxury has undergone a paradigm shift from materialism to experiential value. Millennials seek new and unique curated experiences, to see a place that few or none have seen. Sustainable luxury is the new luxury agenda, it offers the opportunity of a new design meaning for the next generation of megayachts through an holistic approach encompassing a range of issues from addressing pollution to societal contributions. Arctic ice shrinkage is a global concern producing an opportunity for unique luxury experiences and urgently needed scientific research. This project will develop a 300m sustainable luxury Ice-Class Arctic explorer vessel to support scientific research and curated luxury experiences for 36 guests, designed to 13-36 PYC. It will combine an exterior form inspired by architecture with the interior urban flow of a cruise ship resulting in the spectacularisation and conquest of landscape normally associated with cruise ships. Where submarines, diving experiences and drones capture a unique social media narrative of a once in a lifetime experience. The business model is based on private members club such as Quintessentially One. To support the sustainability agenda the vessel and tenders etc. will be hydrogen powered, this hydrogen will be produced by floating wind turbines located around the region.

3. EXTERIOR DESIGN

Contemporary industrial design practices for mood boards involve collecting a series of images. These are a collection of abstract images that enable the designer to communicate emotions, feelings and aspirations. Mood boards are a collection of visual images (e.g. photographs, material samples) gathered together to represent an emotional response to a design brief [18, 19]. This technique enables designers to communicate and express themselves beyond linguistic restrictions. Designers may use this tool to communicate intangible and abstract emotions such as happiness, sadness and calm. Equally, this tool has been employed to enable users to communicate their emotional responses to products, task and their experience through abstract images. The exterior mood board shown in Figure 4, communicates a Retrofuture based minimalistic design.

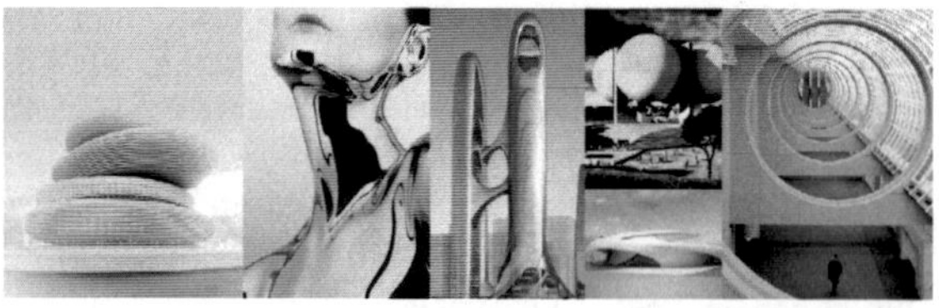

Figure 4: Exterior mood board

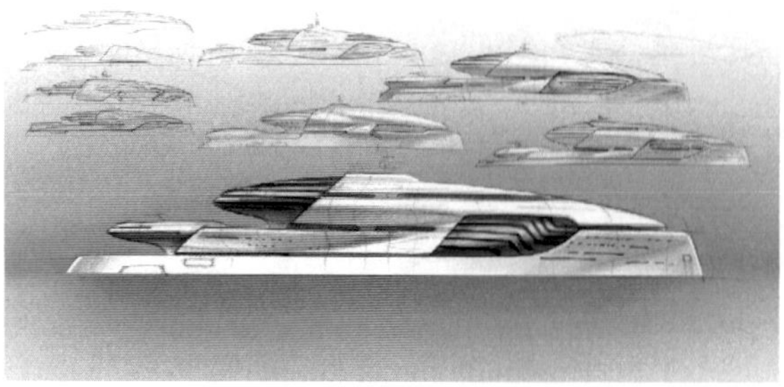

Figure 5: Exterior side profile development

The exterior side profile development of Beluga is shown in Figure 5. Where at 300m Loa, surface disruption techniques are critical to avoid vast expanses of unbroken white surface. The scaling-up of which could result in the creation of amorphous, incoherent and incongruous masses of steel and aluminium. The development of the exterior structure was informed by the surface disruption techniques considered by Nelson [20]. Here the use of material, shape form and colour breaks reduce the perceived visual mass of the superstructure. The horizontal glazing lines act to break up the volume into smaller components. The surface cut lines providing an abrupt step transition from the organic shaped main structure to the fore geometric area of glazing and the aft ellipsoidal shaped geometric structures. The juxtaposition of the fore geometric glazing area and the rear ellipsoidal shaped glazing area, act to visually balance each other, lowering the perceived visual mass of the white main structure. The geometric similarity of the two rear ellipsodial features balance the flowing bow form. The overall streamline flow of the superstructure and the interrelation that it has with the bow give it a dynamic stance and a strong sense of directionality. The exterior form development of the vessel is shown in Figure 6. Here the waisting form of the surface, aft of the fore geometric glazing area, acts to accentuate the sense of streamlining and surface refinement. Forming a gradual surface form transition along the length of the vessel from the square sided bow to the ellipsoidal geometric surfaces on the aft of the vessel. This refinement of the flowing surface form with waisted feature towards the transom is show as a render in Figure 7. The prominence of the ellipsoidal geometric features near the transom are shown as renders from ice level and helicopter view in Figure 8 and 9 respectively.

Figure 6: Exterior form development

Figure 7: Render of vessel in ice

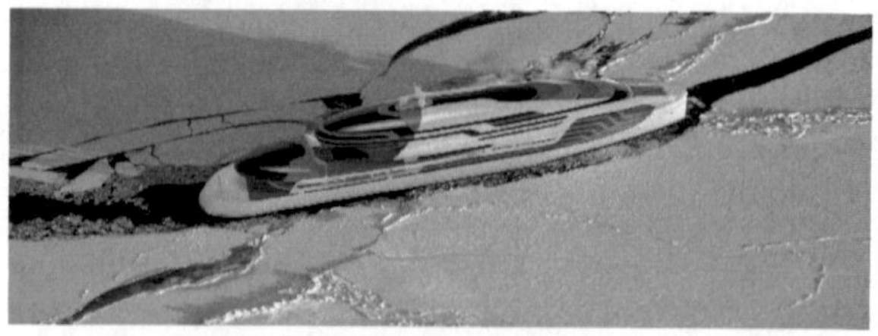

Figure 8: Helicopter view render of vessel

Figure 9: Transom view render of vessel

4. INTERIOR DESIGN

The side profile block layout of MY Beluga is shown in Figure 10. The architectural layout centres on the atrium/main stairwell volume, which has a skylight and acts as a connecting space, bringing together all the functions of the ship. Aft of the atrium on the upper decks there is the guest restaurant with a commanding panoramic view of the environment. Below the guest restaurant is the science and research area, which consists of: laboratories; specimen facilities; learning laboratories for guest; TV studio for documentaries.

Below the atrium are the scientists spaces, which consits of: accommodation for up to 44 scientists and researchers; lounge; dining area; entertainment area. Fore of the atrium volume on the upper decks are the guest entertainment spaces which includes: casino; observation area; sky bar; tennis court. Directly below the guest entertainment spaces are the guest residential spaces which consist of: luxury apartment/ penthouse style suites for up to 36 guests; spa/gym/ pool; guest courtyard. The welcome lounge is located beneath the guest spaces, and beneath the welcome lounge are the crew spaces. The crew spaces consist of: accommodation for 120 crew; crew mess; crew gym; crew lounge. The GA is shown in Figure 11, where the connecting bridges on the various decks of the atrium can be seen. The themes and design concepts of key interior areas will now be delineated and their location on the GA indicated by shading, to aid visual communication of spatial layout.

4.1 INTERIOR SPACE TRANSITION

The critical aspect of the interior design theme is the transition between guest space and crew and scientist space. The interior design theme visual references for this transistion are shown below in Figure 12. The upper images represent guest spaces with strong embellishment of materials and finished, the lower pictures of sci-fi inspired lighting and clinical minimalism, represent the crew & scientist space. The images in between mark the transition process, involving a reduction in material usage and detail, combined with a variation in mood lighting from warm to clinical.

4.2 MAIN ATRIUM

The main atrium is highlighted in the GA shown in Figure 13. It is the connecting space of the vessel, with a skylight. The architectural visual references of the space are shown in Figure 14. Connecting bridges on different levels offers different perspectives and spatial experience to the user. The use of a gradual variation in wood tonality on each deck floor, is used to visually communicate the relative location of the deck within the height of the vessel. This is a planned interaction with the variation of lumination levels provided by the natural lighting through the skylight and the design of artificial light for the space. The tonal variation in veneer colour is shown in Figure 15.

The use of wood veneer on the ceilings and artificial lighting in the atrium, is shown in Figure 16, from the lower deck. Where two spiral staircases and three interconnecting bridges, which join different decks, can be seen. There is also a sofa in the lower deck lounge area, bathed in the natural light from both the side window and the skylight. The view of the atrium from the skylight is shown in Figure 17, here the interesting arrangement of the sky bridges can be seen with their integrated lounge pods. These offer an unique intimate social space with the spectacularisation of landscape. The sofa and table of the suspended lounge pod is shown in Figure 18. It is both a thoroughfare and an intimate space.

Figure 12: Interior space transition

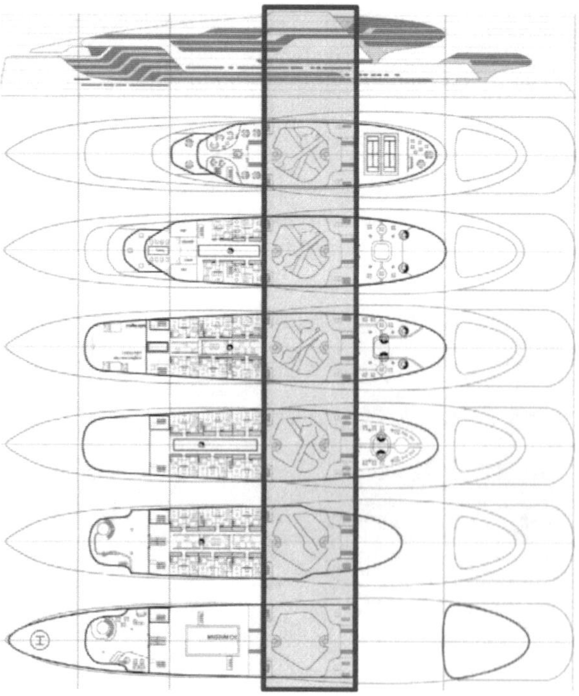

Figure 13: Main atrium highlighted on GA

Figure 14: Atrium architectural visual references

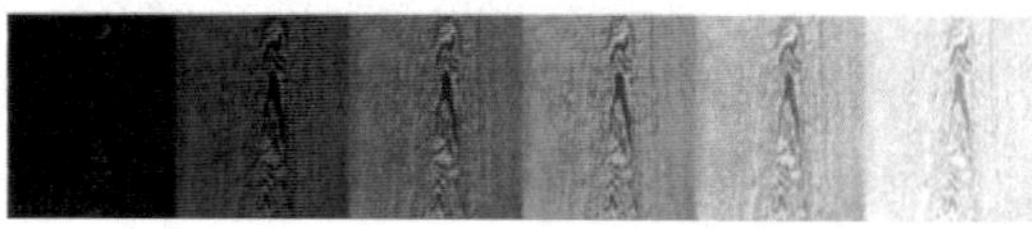

Figure 15: Tonal variation in veneer colour

Figure 16: Atrium view from lower deck

Figure 17: Atrium view from skylight

Figure 18: View of suspended lounge pod

4.3 RECEPTION LOUNGE

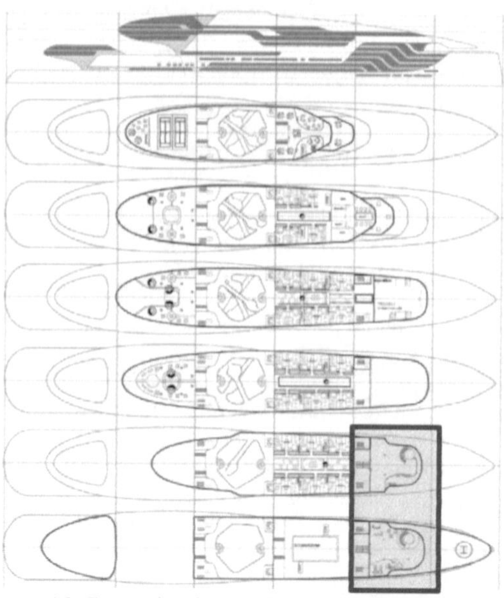

Figure 19: Reception lounge highlighted on GA

The location of the reception lounge is highlighted in Figure 19. This is a mezzanine space, the inspiration mood board images are shown in Figure 20. Here the inspiration is futuristic and geometric.

Figure 20: Mood board of reception lounge

A rendered view of the reception lounge showing the variation in lighting and use of texture, is shown in Figure 21. The tessellated cylindrical surface in the left of the image is the circular staircase and elevator to the next floor. The significant use of glazing bathes the space in panoramic views. The narrow perimeter area on mezzanine deck allows lounge seating areas located there to command a spectacular view without causing significant visual impact on the other seating areas of the lower deck.

Figure 21: Render of reception lounge interior

4.4 RESTAURANT

The location of the restaurant in the GA is highlighted in Figure 22. It has an imposing panoramic view over the three deck levels that it occupies, which can be seen in the side profile view. The visual inspiration images for the restaurant area are shown in Figure 23. They communicate the conquest of landscape experienced by the guests and the sense of flowing space articulated through the use of open and sweeping stairwells. The ideation of the restaurant volume is shown in Figure 24 as asymmetrical decks linked by several spiral staircases. This idea gave way to a symmetrical approach to maintain coherence within the holistic design process of linking interior to exterior. A view of the mezzanine floor of the restaurant from the lower floor of the restaurant, is shown as a render in Figure 25. The use of edge lighting on the floating floors compliments the vast amount of glazing and immersive panoramic views. The geometric pattern on the ceiling and the coherence in style with the floating geometric features compliments the futuristic exterior design language. The use of dark wood flooring and white light coloured ceilings accentuates the sense of space. This is further enhanced by the use of low visual mass furniture. This sense of space and the immersive panoramic views combining to communicate the experiential value of luxury.

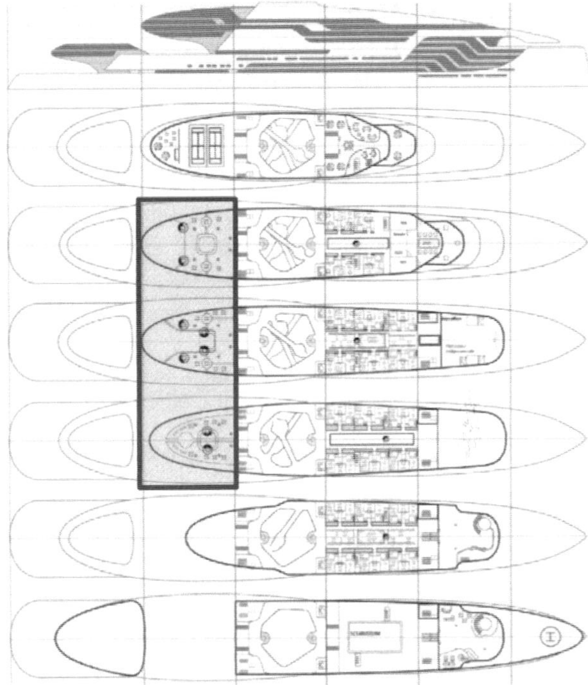

Figure 22: Restaurant highlighted on GA

Figure 23: Restaurant architectural visual references

Figure 24: Ideation of restaurant volume

Figure 25: Render of restaurant

4.5 GUEST COURTYARD

The location of the guest courtyard is indicated on the GA in Figure 26. This is a mezzanine space with several floors connecting the guest apartments to the rest of the ship.

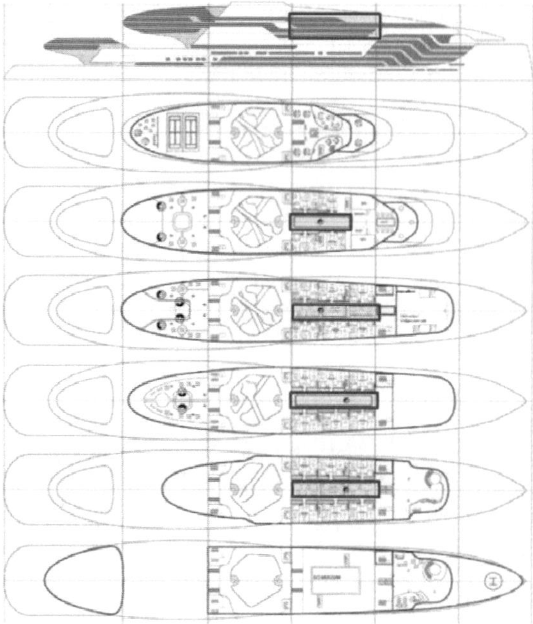

Figure 26: Location of guest courtyard on GA

The architectural visual references of the guest courtyard are shown in Figure 27. These communicate the vastness of space with the drama of lighting in a minimalist interior design theme.

Figure 27: Guest courtyard architectural visual references

Figure 28: Ideation of guest courtyard

The ideation of the guest courtyard is shown in Figure 28. The use of natural materials and an array of poles as an undulating form, breaks up the formal geometry of the space. The sketch development of the hallway and room entrances, with a degree of trim detail development are show in Figure 29. Here the use of colour break with the reflective black door frame and sweeping radius of the wall surfaces make the entrance a focal point within the space. This disruption to the perceived overall length of the wall and hence the space, in terms of proportions, is address through dividing the wall into a series of horizontal feature lines. Whereby the proportions of this array of surfaces acts to accentuate the length of the wall.

Figure 29: Sketch development of room hallway

Figure 30 shows a rendered view of the fore end of the guest courtyard, it has a double height display screen showing the environment outside the ship in real time. Thereby, connecting the guests with their environment. This technology feature acting as a focal point for the fore end of this space, which is complimented by a seating area with table. Combined with the real time exterior view of the display screen, the use of recessed panel lighting and geometric roof beams, convey the idea that natural lighting is reaching the centre of the vessel and the screen a window. The textured panelling on the walls is emphasized by the use of down lighters and lower strip lighting.

The use of biophillic features as space dividers and the undulating array of chrome rods acting as a biomimetic feature, break up the geometric nature of the space to give it the feel of a natural outdoor space. This is shown in Figure 31 and Figure 32, where a series of communal spaces with touch screen media tables enable individuals, couples and groups to engage with and manage their social media. The high bar stool type bench is an informal space for meeting and interacting with new people. The visually light furniture with neutral colours accentuates a sense of space, building upon the minimalistic theme overlayed with biophilia. In the render of the hallway and room entrance, shown in Figure 33, the edge lighting makes the wall appear to float between floor and ceiling. The down lighters can be seen as a pattern of vertical lights drawing attention to the textured surfaces of the horizontal panels. The undulating pattern of chrome tubes gives the hallway both a sense of privacy and spaciousness.

Figure 30: Render of guest courtyard stairs

Figure 31: Render of lower guest courtyard

Figure 32: Render of lower guest courtyard

Figure 33: Render of hallway and room entrance

4.6 GUEST SUITE

The GA of the guest suite is shown in Figure 34, it has an area of approximately 150m^2. Entering from the courtyard the entranced foyer is flanked on both sides by his and her bathrooms. The bathroom to the right has a shower, whereas the bathroom to the left has a bath and a walk in closet with dressing table. In front of the bathrooms facing the outside window are the dining area with work area and bar/lounge area. These flank the super king bed space in the middle, which has a structural wall separating it from the entrance foyer and room dividers with slats on either side. The mood board for the guest suite is shown in Figure 35. The ideation of the space and furniture is shown in Figure 36. With the detail design of the sink unit and bed in Figure 37.

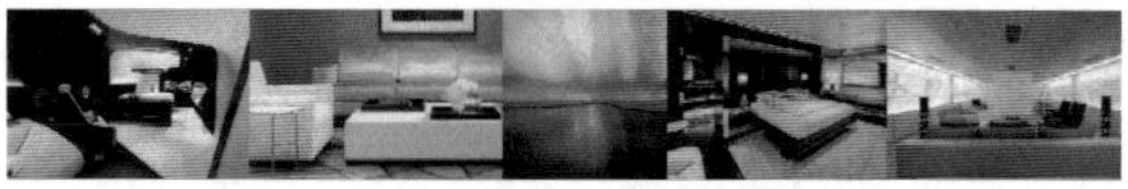

Figure 35: Mood board

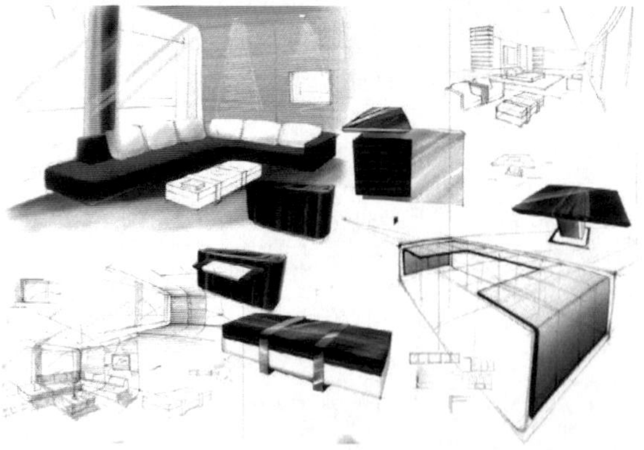

Figure 36: Ideation of guest suite

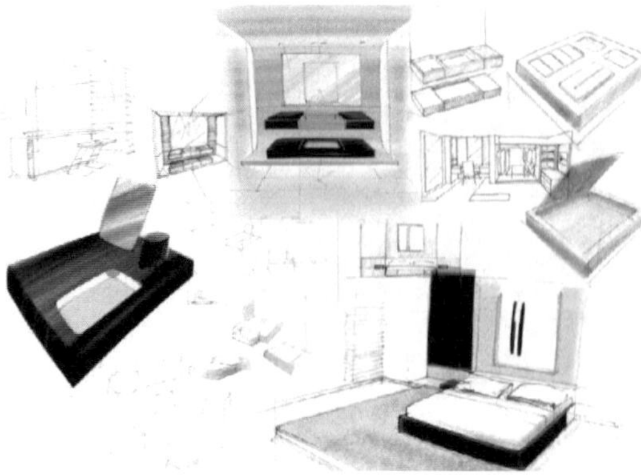

Figure 37: Detail design of sink and bed

Figure 38 shows the camera views of the interior renders show in Figure 39-42. Camera location A is shown in Figure 39 showing the bar/ lounge area. The use of surface texture on the walls is accented by the use of down lighters. Camera location B is shown in Figure 40, showing the panoramic view from the work area which overlooks the dining area. The visual balance of wood floors and neutral ceiling with a corner radius, combined with the use of recess lighting emphasizes the sense of space.

Camera location C in Figure 41, shows the attention to detail applied to the bath area with down lighters emphasising the use of texture on the back wall and the use of edge lighting to frame the bath as an object in space. The bathroom sink in camera location D is shown in the render in Figure 42. Here the use of edge lighting delineates the sink as a space, the contrast of colour and materials makes the feature a focal point. The large mirror increases the perceived volume of the room. Where the band of light around the users inner mirror area, connect the user with the mirror function without the distraction of the overall size of the feature.

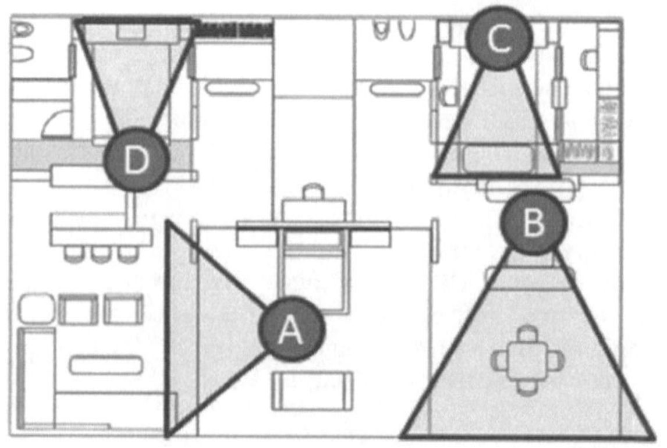

Figure 38: Render camera positions

Figure 39: Render of camera view A

Figure 40: Render of camera view B

Figure 41: Render of camera view C

Figure 42: Render of camera view D

4.7 LABORATORIES

The location of the laboratories on the GA are highlighted in Figure 43. It is a large volume atrium space with three decks of laboratories around the perimeter. Where the glass walled laboratories on each deck level enable guests to observe the scientists from the atrium, which acts as a social space and an interface between the guests and the scientists. The architectural visual references, shown in Figure 44, communicate a clinical minimalist approach with accent lighting. A perspective view of the atrium, the small group meeting area and stairwell are shown in Figure 45, as a basis for the spatial exploration of these key elements. The detail development of the stairwell as a biophillic design element, with strip lighting to give a sense of moving light and mood is shown in Figure 46. The design detail of the collaborative breakout spaces and lighting is shown in Figure 47, with presentations screens and movable furniture. These spaces are used for both collaborative research and as meeting space for scientist to engage guests in informal discussion. There is a large lecture theatre and media studio for formal lectures and social media content creation. A laboratory space is shown in Figure 48, where a novel use of an irregular pattern of horizontal lighting strips breaks up the volume of both the workbenches and the overall space, creating each laboratory as a lighting feature to be view from the atrium space.

Figure 44: Laboratories architectural visual references

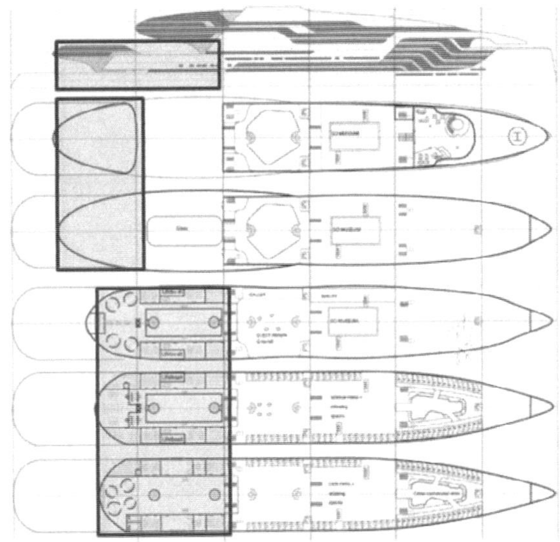

Figure 43: laboratories highlighted on GA

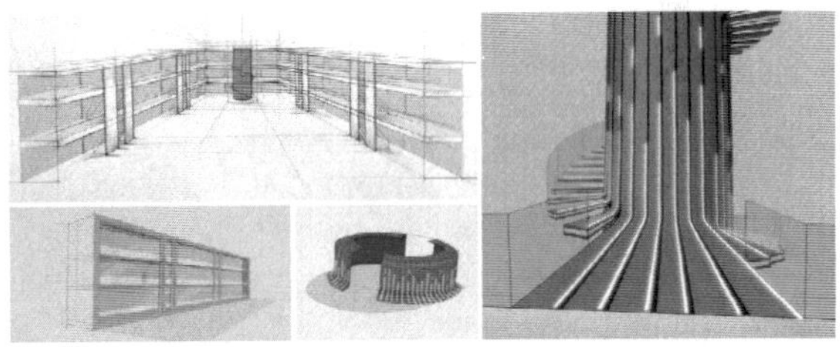

Figure 45: Development of atrium, meeting area and stairwell

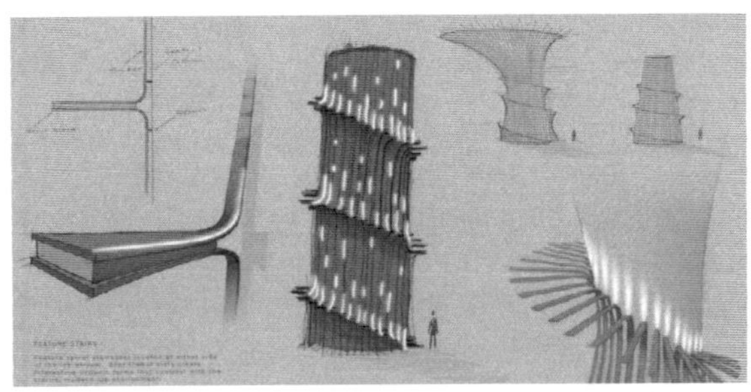

Figure 46: Biophillic detail development of stairwell

Figure 47: Collaborative breakout spaces

Figure 48: Lighting detail of laboratory space

4.8 CREW AREA

The location of the crew area in the GA is highlighted in Figure 49 on the lower decks at the bow of the ship. Here there is a connected central volume of communal space with connects by bridges to the crew individual rooms which are individual spaces around the periphery of the vessel. The inspiration images are shown in Figure 50, they show the contemporary architecture of and pictures of the artic environment. The ideation of the crew area based on a central communal hub for dining and socialising is shown in Figure 51. The use of transparency and the enclosed volume of a space within a space, which is inspired by the glacier image, makes playful use of the perceived sense of volume.

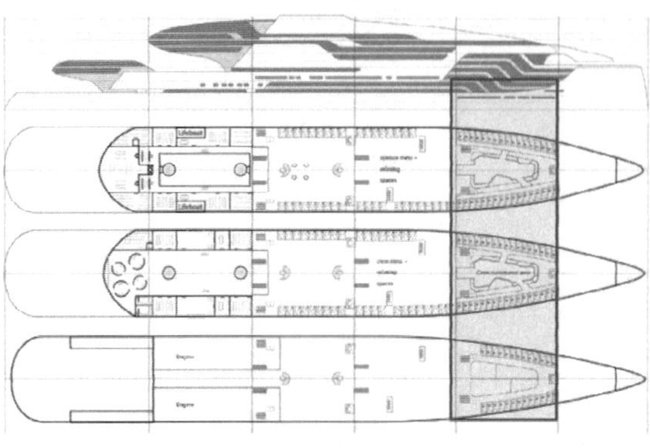

Figure 49: Location of crew area on GA

Figure 50: Crew area visual references

The use of transparency and strip lighting to create a sense of atmosphere and space is shown in the renders of the area in Figure 52. Here the use of edge lighting and highly reflective surface act to convey a greater sense of space. The large lighting features relate to the bridges of the living hub to create a sense of community. It has a Scandinavian minimalist style.

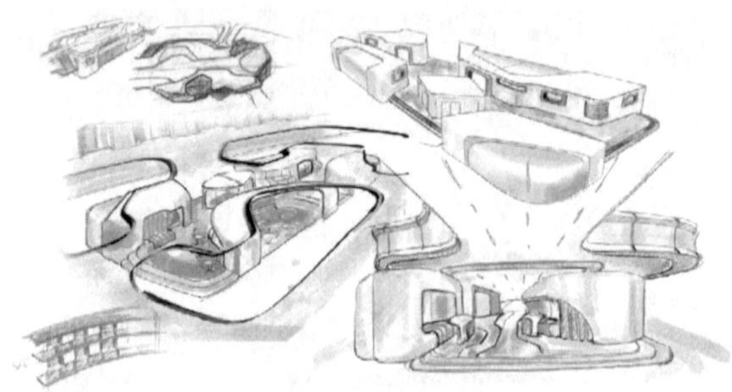

Figure 51: Ideation of crew area

Figure 52: Renders of crew area

5. DISCUSSION

The 'Beluga' Arctic explorer gigayacht concept presented in this paper is a Design-Driven Innovation proposal, to create a new design meaning of a gigayacht as a sustainable luxury experience for the new millennial market sector. The highly stylised exterior form and the strong influence of avant-garde architecture on the interior, reflected in the large interior volumes communicating space as statement of luxury are an implementation of the new design philosophy of gigayacht design proposed by McCartan et al [1]. This is essentially an unconstrained approach to creating unique and design research informed user experience, to radically innovate a new design meaning for millennials as the future customer of the industry sector. This differs from the Future Feadship concepts reviewed by McCartan et al [21], which are fully resolved technical designs.

The 'Beluga' concept has not observed the architectural layout restrictions of exhaust systems and ventilation systems on the centre-line of the ship. Nor does it implement the PYC 13-36 code for fire regulations in terms of MVZ and means of escape. It is in essence a floating interior design proposal based on benchmarked hull from a 300m passenger ship for key dimensions. The benefit of PYC 13-36 in this design proposal is the ability to have a wider use of natural materials in terms of fire safety regulations compared to full SOLAS regulations of a passenger ship. The objective of this concept is to communicate the user experience that this radical innovation offers. The next stage of the design process will be the full structural development of the vessel and the implementation of PYC 13-36 regulations. This process will create a dialogue between Marine Designers and Naval Architects based on the elucidation of a business model of the vessel operation. The outcome of which will be a design compromise between conventional design regulation implementation and the need for alternative design methods to achieve the expectation of the millennials user experience.

An understanding of the needs and aspirations of millennials is critical in developing the business model for this vessel. This has been delineated in the introduction. They are aware the every decision makes an impact, when it comes to investing and consuming. Millennials also want to be stewards of the environment; protecting it, and bringing ecosystems back to full-strength. This is their rationale for applying the '*Triple Bottom Line'* an accounting framework that takes into account social, environmental, and economic performances instead of simply focusing on financial performance. The 'Beluga' Arctic explorer gigayacht concept is predicated on facilitating these user needs through addressing the sustainable luxury agenda. It does so by facilitating critical research for the future well being of the environment in a zero emission vessel powered by hydrogen sourced from renewable energy. Where the vessel is designed a series of unique user experiences, a continuous passengers' flux along the ship in an urban-like environment, where different design languages meet. The use of technology to engage their predilection for social media results in a vessel which is a connected space, with the use of drones and submarines to produce striking social media content.

The proposed business model addresses the market sector between superyachts and ultra-luxury cruise ships identified by McCartan and Edens [22]. It is based on a floating private members club, such as Quintessentially One, which purports to be the World's finest floating private club [23]. It is a 220m superyacht which operates as a floating destination, with 12 triplex apartments located on the top decks. They are available for purchase on a leasehold basis for the life of the ship, with prices ranging from 8m Euro to 12m Euro. It has 12decks, a residential area of 13,000m^2 and a common area of 9,300m ^{2} supported by 490 crew. The amenities include: club reception; beach club; night bar; marina; business centre; library; kids area; bespoke services; spa; emporium [23]. A critical aspect of the business model, which will be developed in further work, is the cost of developing a network of far shore wind turbines and electrolysis systems to produce and store hydrogen. This renewable energy network could also be used to support other industries in the region, thereby supporting the development of more eco-friendly products.

The radical exterior form language is a significant departure from both the conservative layer cake approach of traditional superyachts and the box volume of cruiseships. It aims to engage millennials in a visceral response, to an embodiment of technology for an optimistic future based on Retrofuturism. This visual style incorporates various aspects of the Streamline Moderene and Art Deco architectural style, giving the ship superstructure the appearance of a space ship in a science fiction environment. Retrofuturism can be summarized as the future as seen from the past. It is directly inspired by the imagined future which existed in the minds of writers, artists, and filmmakers in the pre-1960 period who attempted to predict the future. The German architecture critic Niklas Maak [24], sees retrofuturism as nothing more than an aesthetic feedback loop recalling a lost belief in progress, the old images of the once radically new.

A common thread in Retrofuturism is dissatisfaction or discomfort with the present, this aligns with millennials environmental concerns, to which it provides a nostalgic contrast. Retrofuturism suggests a world which may be more comfortable or at least more capable of being understood while evoking a nostalgia for a time of forward-looking hope and romance. [25] The idea of "Beluga" as a vessel applying science for the benefit of society and the environment, while being a zero emission vessel embraces these idealisms. This radical innovation of design meaning in addition to pure nostalgia, may act as a reminder of the forgotten ideals of technology benefitting mankind.

The interior engages in the architectural approach of space as luxury and the conquest of landscape, using significant amounts of panoramic glazing on the upper decks. The use of lighting, texture, materials and biophilia, facilitate the interior space transition between guest space and crew and scientist space. Where guest spaces have strong embellishment of materials and finished. In contrast to this the crew & scientist space have sci-fi inspired lighting and clinical minimalism. The transition process between the two involves a reduction in material usage and detail, combined with a variation in mood lighting from warm to clinical.

Tschumi [26] expanded his architectural concerns to address the issue of cities and their making. He suggests that architecture can accelerate the events of everyday life thorough new forms of organization. These have been considered in this proposal, through the complexities of the architectural process and the rich texture of events that define urban reality in the context of the application of this new design philisopy implemented through Design-Driven Innovation. The significant volume of the science area atrium space in Figure 45 is an unresolved design proposal. It would benefit from the implementation of walking bridges, to engage in spectactuarlisation, as used in the main atrium with the sky bridges shown in Figure 17. The bridges in the atrium are playful elements that enables the guest to experience the open space in different perspectives. These different perspectives virtually enlarging the spatial experience. As example of this element of play, we refer to Pompeia Social Centre in São Paulo by Lina Bo Bardi [27], which is designed for citizens to play and meeting. The use of the bridges to connect different deck levels with the laboratory atrium space, adds to the sense of an urban-like environment.

Figure 53: Bridge view of Pompeia social centre [27]

Streets are a predominant element in cities as promenades are in cruise ships. Defined as paths, conduits, structural lines, and even a place with its own rights, the design of streets is still an important issue to be understood and resolved in order to improve the residents' quality of life or passengers' entertainment experience in the case of a cruise ship. Stress is a public health concern that affects residents and is constantly originated by multiple circumstances. This problem not only affects mental health but also overall well-being by reducing the capacity of people to restore themselves on a daily basis .Environmental psychologists show that people can release stress and mental fatigue by providing cities with a restorative environment. Urban environments provided with natural light, vegetation, water and a certain amount of wilderness, increase the psychological restoration of people. Biophilic design, aimed at maintaining, enhancing, and restoring the beneficial experience of nature in the built environment, allows people to be part of the larger whole. Streets and promenades can be designed as places that improve people's well-being on a daily basis. [28]

Hidalgo [28] proposed that the connection between environmental psychology and urban design was in need of further development and presented theoretical relationships among urban design theory on streets, environmental psychology and the restoration theory, and biophilic design. Where these interrelations were inputs to define guidelines for the future design of restorative streets, which could be equally applied to promenades on cruise ships as urban spaces. Stress affects health by threatening the immune system, producing psychological problems it also challenges well-being when a person's capacity for adaptation of their environment is not reached. Stress and mental fatigue can be released by psychological restoration. The capacity of people to recover their health status from stress in urban environments should be a major concern in urban planning practices; it should be a main concern in streets design. In order to face stress-related outcomes, it is necessary to have interdisciplinary approaches. Hidalgo [28] referred to the attention restoration theory (ART) and the stress restoration theory (SRT).

Both theories, provide evidence of how some environmental resources improve people's psychological health. For instance, several studies show that the use of vegetation, natural water and green landscapes helps in mental health recovery. Specific urban places such as parks, museums, spiritual temples, and healing buildings also provide restorative experiences. Most of the elements that constitute these spaces include nature and the possibility of personalizing the environment. From another perspective, biophilic design focuses on restorative environments that enhance the innate urge of people to affiliate with nature. This approach, originated in biological sciences. The following seven categories summarize the biophilic elements: organic forms and structure, day lighting, natural ventilation, natural sounds, palette of colours, mimicking landscape, and bio-inspiration gardens. This biophillic design approach has been implemented within the guest courtyard. Through an innovative use of artificial lighting and large screen technology, in addition to the biophillic and biomimetic features.

Creating a 'wow' visceral response to visual first impression is becoming of secondary importance to the implementation of texture as a primary objective in superyacht interior design. A third skin of fine textures layered around the client or guest's second skin is beginning to define the luxury yacht experience. One of the epicentres of the texture trend is Italy. Fabric is a particularly critical factor on furniture. People at home wear clothes, whereas, on a superyacht they may be wearing next to nothing. On this basis Stefano Pastrovich proposes that interior designers need to consider how the fabric makes direct contact with the body. As peeling oneself off leather or a glossy hard-finished chair can feel uncomfortably sticky. He tends to reduce the amount of visual information in his interiors, but where an accent is needed to create a focal point he often uses rich coloured, highly textured fabric. Proposing that because fabric advances to the eye more than a painted surface, less can indeed be more. Building upon these considerations the interior design proposal has a strong use of natural fabrics to offer breathability combined with a sensual experience of textural detail such as the sofas cushions and chairs seats.[29]

Antonio Romano of Studio Hotlab believes that the superyacht industry has a long way to go before it arrives at the level of a detail achieved by luxury automotive brands such as Bentley, but it is a goal that the superyacht industry should aim for. An example of which is the attention to detail with which they match wood veneers with leather. In terms of the luxury experience of the interior he proposes the texture will do the talking, with guests first becoming comfortable and them over a period of time experiencing texture and noticing detail. [29] This experiential value of texture has been fully considered in the development of the guest suite, shown in Figure 38 to Figure 42. This informed approach makes use of surface texture on walls with accent lighting to accentuate the texture effect. Where the variation in perceived scale of texture associated with viewing distances offers a unique user experience, in addition to tactile experience of texture on the furniture.

6. CONCLUSIONS

Millennials are the new sustainable luxury customer, they seek experiences not products. Sustainable luxury is an opportunity for a significant paradigm shift in the business model of the superyacht industry, combining luxury travel with philanthropy. A new business model between superyacht charter and high-end cruising will fund larger vessels and infrastructure. Moving away from individual charter and ownership to more of a high-end luxury crusing approach. A deep interrogation of the business model is required to elucidate the potential of this Design-Driven Innovation proposal. Further dialogue with stakeholders will be carried out by the authors to elucidate the emotional response of millenials to the radical innovation in exterior design language and user experience.

7. ACKNOWLEDGEMENTS

The authors wish to thank the following 3rd year Coventry University Boat Design students, who's professional collaborative EBDIG-IRC project constitutes the visualization work presented in this paper: Jae-Hoon Ahn; Michael Ashley; Youn Chae; Minsoo Kim; James Marshall; Min Joo Kim.

8. REFERENCES

1. MCCARTAN, S., THOMPSON, T., MORI, L., BROSSA, A., and VERHEIJDEN, B., 'Design-driven innovation: a new design meaning for superyachts as a less egocentric experience', RINA Conference: Marine Design, 2-3 September 2015, London, UK.

2. REIN, G., http://wwd.com/fashion-news/designer-luxury/millennials-luxury-spending-10417737/

3. https://www.digitalagent.com/triple-bottom-line-millennials/

4. https://www.morganstanley.com/press-releases/ultra-wealthy-millennials-are-idealistic-but-not-alienated-challenge-stereotype-of-the-idle-rich_f86c1c36-37eb-4015-b948-f35323743a35

5. GONZALEZ, F., 'Sustainability and the Luxury Industry', LuxurySociety, 3rd November 2009 available at http://luxurysociety.com/articles/2009/11/sustainability-and-the-luxury-industry accessed 30/03/15

6. FUAD-LUKE, A., 'Design Activism: Beautiful Strangeness for a Sustainable World', ISBN: 978-1844076451, Routledge, 2009.

7. THE FUTURE OF LUXURY, 'TFOL Talks To: Diana Verde Nieto, Founder of Positive Luxury', 26th September 2013, available at http://www.thefutureofluxury.co.uk/tfol-talks-to-diana-verde-nieto-founder-of-positive-luxury/

8. http://www.bulgarihotels.com/en-us/london/community/green-initiatives

9. BRE,' Code for a Sustainable Built Environment', available at http://www.breeam.org/page.jsp?id=347

10. THE FUTURE LABORATORY, 'Luxury futures: A global snapshot of new and emerging trends in the luxury travel market', International luxury travel market, Cannes, 3-6 December, 2012

11. http://www.bsr.org/en/collaboration/groups/sustainable-luxury-working-group

12. http://arp.arctic.ac.uk/challenge/

13. http://www.enr.gov.nt.ca/en/state-environment/73-trends-shipping-northwest-passage-and-beaufort-sea

14. http://news.nationalgeographic.com/2016/08/crystal-serenity-luxury-cruise-arctic-northwest-passage/

15. http://www.cruisecritic.co.uk/articles.cfm?ID=2662

16. WEIGHTMAN, D., and MCDONAGH, D.,'People are Doing it for Themselves', in the Proceedings of the International Conference on Designing Pleasurable Products and Interfaces Forlizzi, J, Hanington, B and Ayoob, E (eds) ACM Press, Pittsburgh, Pennsylvania,USA p34-396, 2003.

17. VERGANTI, R., Design-driven innovation: changingthe rules by radically innovating what things mean,Harvard Business School Publishing Corporation, ISBN978-1-4221-2482-6, 2009.

18. MCDONAGH, D., STORER, I., 'Mood boards as adesign catalyst and resource: researching an under-researched area', Design Journal 7 (3): 16-31, 2005.

19. MCDONAGH, D., DENTON, H., 'Exploring thedegree to which individual students share a common perception of specific mood boards: observations relatingto teaching, learning and team-based design', Design Studies 26 (1): 35-53, 2005

20. NELSON, T., 'The potential for surface disruption techniques in the aesthetic evolution of superyachts', RINA Conference: Marine Design, 3-4 September 2014, Coventry University, Coventry, UK.

21. MCCARTAN, S., STUBBS, E., CREA, N., ROWLES, D., VERHEIJDEN, B., and SHOUTEN, R., 'Design-Driven Innovation: Avant garde 50m superyacht under 500GT', RINA Conference: RINA Conference: Design and Construction of Super & Mega Yachts, 13-14 May 2015, Genoa, Italy.

22. MCCARTAN, S., and EDENS, J., 'Design-Driven Innovation: A new luxury maritime sector between cruising and superyacht charter', RINA Conference: RINA Conference: Super & Mega Yachts, 8-9 May 2013, Genoa, Italy.

23. QUINTESSENTIALLY ONE, "The World's FinestFloating Private Club", http://www.quintessentiallyone.com/, Accessed05/04/13

24. MAAK, N., 'Goodbye Retro-Futurism · A farewell to our perpetual nostalgia for the future'. 032c Culture Magazine, Issue 9, p117, Berlin, 2005.

25. LATHAM, R., 'The Oxford Handbook of Science Fiction', 2014. ISBN 9780199838844

26. TSCHUMI, B., 'Event-Cities 2', MIT Press, 2001. ISBN 0262700743

27. http://www.archdaily.com/575429/spotlight-lina-bo-bardi

28. HIDALGO, A.K.,' Urban Streets: Towards a Psychological Restorative Function', 2nd Future of Places International Conference on Public Space and Placemaking, Part II: Streets as Public Spaces and Drivers of Urban Prosperity, Buenos Aires, Argentina, 1-3 September 2014.

29. POSTEL, L.,' Importance of texture in yacht interior design', ShowBoats International, Issue 62, October 2012.

9. AUTHORS BIOGRAPHY

Dr Sean McCartan holds the current position of Course Tutor, Boat Design at Coventry University, UK. His key research area is TOI (Transfer of Innovation) from other sectors to the marine industry, in the areas of Design-Driven Innovation (DDI), advanced visualisation and Human Systems Integration(HSI). He leads the EBDIG (European Boat Design Innovation Group) network, which includes Chalmers University; Genoa University; TU-Delft; and a number of leading European marine design consultancies. He is currently project co-ordinator for the Leonardo TOI project EBDIG-WFSV (European Boat Design Innovation Group - Wind Farm Support Vessels), which aims to develop online training material for Naval Architects in the subject areas of: Human Factors; WFSV design (Industrial Design); WFSV mothership design (Industrial Design).

Alan Barrett holds the current position of Course Director, Transport Design, in the Department of Industrial Design at Coventry University. He has extensive experience as both an automotive designer and a superyacht designer.

Lawson Swinfield holds the current position of lecturer in Automotive and Transport Design, in the Department of Industrial Design at Coventry University. Before joining Coventry University he was a Principal Designer at Jaguar Land Rover for over 12 years. Where he was Principal Designer for XK8, XKR, XJ8, XJ, XK150, XF, XJ350, special editions and sport performance design concepts. He was also design manager for Accessories.

Bob Verheijden holds the current position of Head of Department Design, Academy Minerva, Groningen, NL He has over 20 years professional practice in the following disciplines: architecture; urban planning; interior design; product design; multimedia; brand development. He is a Member of the BKVB (The Netherlands Foundation for Visual Arts, Design and Architecture), and has won a number of prestigious awards for his architectural and furniture designs.

Prof. Dario Boote holds the current position of Ship Structure Professor at the Naval Architecture section of the Department of Electrical, Electronic, Telecommunications Engineering and Naval Architecture (DITEN) of the University of Genoa. He is the Chairman of the Bachelor and Master Course in Yacht Design in La Spezia. His initial experiences include a long research activity in the field of Ship and Offshore Structures followed, since 2000, by an intense activity in the field of sailing and motor yachts. From 2009 to 2012 he has been Chairman of the V.8 ISSC Committee on "Yacht Design".

Luca Mori holds the current position of Senior Architect at De Jorio Design International, Cruise Ships Interiors Designer. Where for over 15 years he has practiced interior design for luxury cruise ships and cruise ferries public areas (shopping galleries, restaurants, bars, cafeterias, coffee shops, discotheques, show lounges), offices and private houses. This senior role has involved the following responsibilities: General Arrangements optimization in respect of marketing strategies and ship-owner requests; ship owner support in materials and finishings choice; architectural concept and ship refurbishment consulting; coordination of yard and contractors; supervision on site of quality and design conformity; teaching and lecturing in vessel interior design workshops at both Plitecnico di Milano and IUAV- Venice. He holds a Degree in Architecture from Genoa University.

Ronno Schouten holds the current position of manager design at Studio De Voogt. He is responsible for this department of Feadship and the design work.

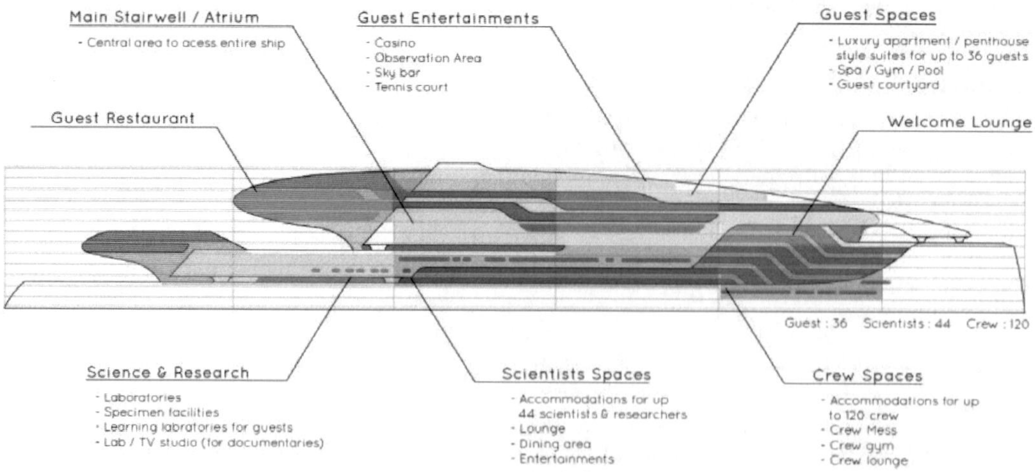

Figure 10: Side profile of Beluga indicating key areas

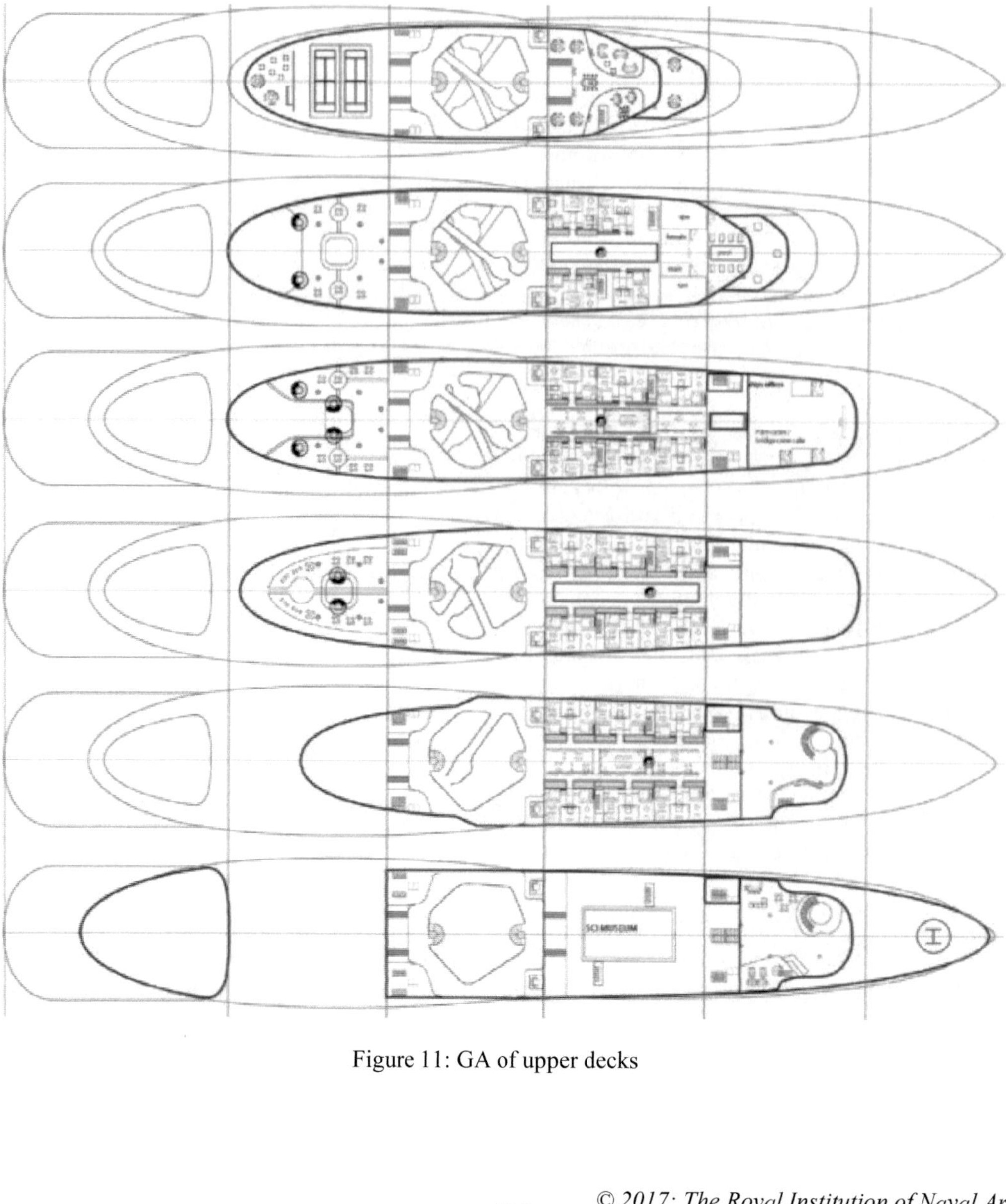

Figure 11: GA of upper decks

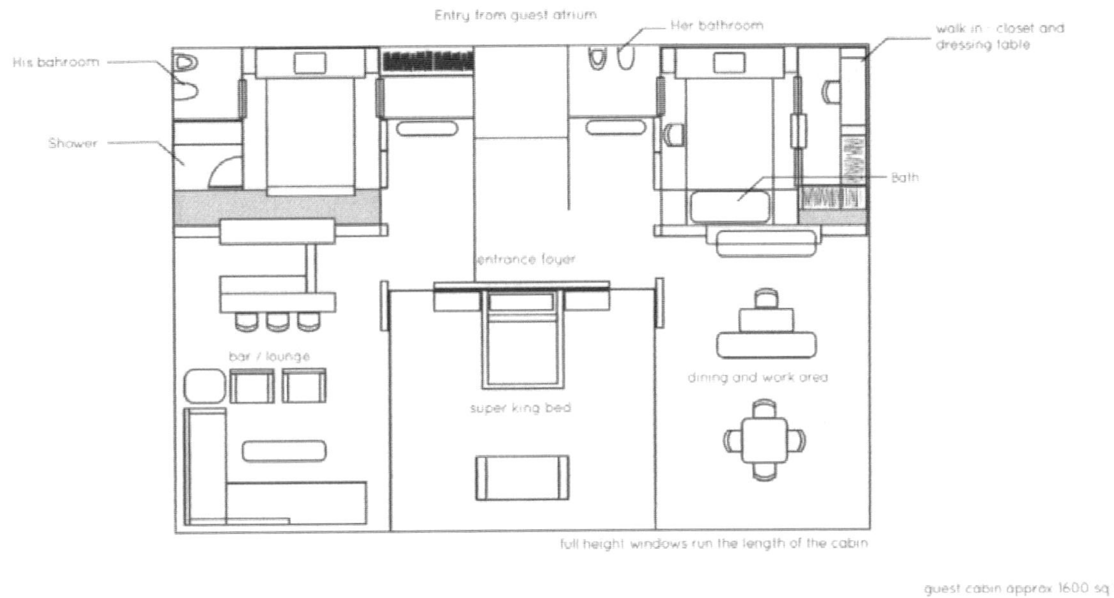

Figure 34: GA of guest suite

AUTHOR'S CONTACT DETAILS

Antonios-Vassilios Lalechos
LePlan
lalechos@leplan.gr

Carlo Russo
Lloyd's Register EMEA
Carlo.Russo@lr.org

Dario Boote
University of Genoa
dario.boote@unige.it

Francisco Miguel Montero
MARIN
f.miguelmontero@marin.nl

Mark Beavis
ACO Marine
pwheater@seabornecomms.com

Nicola Petacco
University of Genoa
nicola.petacco@edu.unige.it

Stefano Brizzolara
Virginia Tech
stebriz@vt.edu

Tatiana Pias
University of Genoa
tatianapais@hotmail.it

Carlo Bertorello
University of Naples Federico II
bertorel@unina.it

Cristina Barba Fau
Lloyd's Register EMEA
cristina.barba-fau@lr.org

Ed Harland
Houlder
Ed.Harland@houlderltd.com

Kimmo Kokkila
ABB
kimmo.kokkila@fi.abb.com

Maurizio De Giacomo
Aviomarine Srl
aviomarinesrl@gmail.com

Robert Shaw
Unitec Institute of Technology
rshaw@Unitec.ac.nz

Stefano Gaggero
University of Genoa
stefano.gaggero@unige.it

Ugur Mutlu
University of Southampton
um1r14@soton.ac.uk